T0146055

Mapping
Disease
Transmission
Risk

MAPPING DISEASE TRANSMISSION RISK

Enriching Models Using Biogeography and Ecology

A. Townsend Peterson

JOHNS HOPKINS UNIVERSITY PRESS
BALTIMORE

© 2014 Johns Hopkins University Press
All rights reserved. Published 2014
Printed in the United States of America on acid-free paper
9 8 7 6 5 4 3 2 1

Johns Hopkins University Press
2715 North Charles Street
Baltimore, Maryland 21218-4363
www.press.jhu.edu

Library of Congress Cataloging-in-Publication Data

Peterson, A. Townsend (Andrew Townsend), 1964– author.
   Mapping disease transmission risk : enriching models using biogeography
and ecology / A. Townsend Peterson.
      pages cm
   Includes bibliographical references and index.
   ISBN 978-1-4214-1473-7 (hardcover : alk. paper) —
ISBN 1-4214-1473-2 (hardcover : alk. paper) — ISBN 978-1-4214-1474-4
(electronic) — ISBN 1-4214-1474-0 (electronic) 1. Medical mapping.
2. Medical geography.   3. Public health surveillance.   4. Epidemiology.
I. Title.
   RA792.5.P48 2014
   614.4´2—dc23   2013046568

A catalog record for this book is available from the British Library.

Special discounts are available for bulk purchases of this book.
For more information, please contact Special Sales at 410-516-6936
or specialsales@press.jhu.edu.

Johns Hopkins University Press uses environmentally friendly book
materials, including recycled text paper that is composed of at least
30 percent post-consumer waste, whenever possible.

Contents

Preface

This book was born from a combination of my long-term fascination with diseases and the intricacies of their transmission and my exploration of new and interesting methods in biodiversity science, biogeography, and ecology. In working toward this synthesis, I frequently found myself being the only evolutionary biologist among public health experts, or the only person interested in disease biology among biodiversity scientists. As a consequence, this volume had an extremely interdisciplinary birth—it falls at the little-explored juncture between public health, biogeography, and ecology.

The biodiversity world received an embarrassment of riches with the development of large-scale biodiversity databases in the late 1990s. Biodiversity scientists found themselves inundated with millions of primary biodiversity data records, but without robust and effective tools in hand with which to analyze such rich information. The past 20 years have seen explorations of such tools: inventory statistics (Colwell and Coddington 1994), place-prioritization algorithms (P. Williams et al. 1996; Sarkar et al. 2006), and ecological niche modeling and species distribution modeling algorithms (Peterson et al. 2011). These tools have led biodiversity science to several exciting new insights and understandings, although much future work remains (Yesson et al. 2007; Peterson et al. 2010).

In the public health world, on the other hand, mapping disease occurrences (and, by extension, the risk of transmission) has been a challenge and point of interest for centuries (T. Koch 2005). Many disease atlases and books compiling and presenting methods for such mapping exercises have been published (Keeling and Rohani 2008; Pfeiffer et al. 2008; Lai et al. 2009; Hay et al. 2013), yet the field remains incomplete, for reasons elaborated in this book. Basically, the challenge is to develop those maps in the context of biogeography and ecology—the two branches of biodiversity science that are relevant to these questions. In general, their focus is on the question of why species are where they are, and why they are not where they are not, albeit from very different points of view.

Many of the current disease mapping methods in public health and epidemiology are not based on a biogeographic and ecological context, a lesson that is only beginning to be appreciated in public health and epidemiology circles (Brooker et al. 2002). Far too frequently, disease occurrences are treated as a spatial phenomenon only. Various adjustments are made to take sampling, human or animal populations, and aggregation areas into account, but simple surfaces are fit to summarize and interpolate disease transmission rates at sites falling in between sites of known disease occurrence. Perhaps as a result of a general focus on chronic disease, as opposed to infectious disease, environments are treated as an exposure, more than as a context. As a consequence, when environmental variation is taken into account in such analyses, too often the treatment is cursory, or not cast in appropriate contexts of ecology and biogeography.

In reality, however, disease cases are simply occurrences of species (the virus or bacterium or fungus or whatever the pathogen is). Ecologists have long known that distributions of species are structured not by the place per se, but rather by the conditions manifested at that place (Grinnell 1917, 1924; Hutchinson 1957, 1978). Under this view, species' distributions are rarely even, simple, or entire in shape, but rather tend to be complex, fragmented, and highly irregular. Only by understanding the underlying environmental drivers that translate into the spatial drivers can one hope to reconstruct and map this complexity effectively.

My interest in disease transmission risk mapping dates back decades, thanks to a sister who was working with vector-borne diseases. Frequently, perhaps to my parents' dismay, dinnertime conversations swung to how rabies is transmitted or to some other issue that did not mesh well with a relaxed meal. Indeed, perhaps my first involvement in research efforts was in trapping birds for serological monitoring during the large St. Louis encephalitis outbreak in the 1970s. Although my academic training took me far from the fields of public health and epidemiology, the ideas and inspiration were there. As a result, when given the opportunity by generous colleagues at the U.S. Centers for Disease Control and Prevention and at institutions in Brazil and Mexico, I was excited to explore the implications of lessons learned in the biodiversity world in addressing this challenge.

In sum, this book sets out to provide a framework for recasting disease transmission risk mapping in an appropriate conceptual context of biogeography and ecology. This volume is neither a manual nor a how-to book; furthermore, it is not a comprehensive review of spatial epidemiology or a full presentation of ecological niche modeling (e.g., Peterson et al. 2011). I extend my apologies in advance to anyone whose work I did not cite, but should have—this book was never conceived as a full review of either of these sets of literature. Rather, it is a conceptual guide designed to equip its readers with the appropriate set of thinking tools for characterizing, analyzing, and understanding why a particular disease is transmitted where it is and not elsewhere. Because this volume presents a particular methodological viewpoint, I have emphasized examples that follow this means of approaching disease transmission risk mapping. I hope that readers will therefore forgive a rather high rate of citation of my own work, which does not reflect ego but rather offers detailed illustrations of the points in the book. My hope is that, with a rigorous, conceptually founded understanding in hand, readers can develop their own suites of analyses tailored to their particular challenges and needs.

Acknowledgments

Many, many people have contributed to the genesis and development of this book. Perhaps my acknowledgment of their contributions is deeper than might be customary, as my training is not in public health, epidemiology, microbiology, medicine, or anything of the sort. As a consequence, every study and entrée that I have made into this field has been thanks to patient instruction and counseling by someone who does have that training. I have attempted to list all of these valued colleagues below; if I have forgotten some, it is not out of any deliberate omission.

At the U.S. Centers for Disease Control and Prevention, my valued and much appreciated colleagues include Mark Benedict, Ben Beard, Darin Carroll, Ken Gage, Nick Komar, Roger Nasci, Jim Mills, and Joel Montgomery. In other settings, similarly esteemed colleagues include Alzira Almeida, Dan Bausch, Sinval Brandão-Filho, Jane Costa, Elisa Cupolillo, Rita Donalisio, Des Foley, Camila González, Rodrigo Gurgel, Karl Johnson, Herwig Leirs, Rebecca Levine, Carmen Martínez-Campos, Janine Ramsey, Victor Sánchez-Cordero, Jeffrey Shaw, Dave Wagner, Lance Waller Rick Wilkerson, and Xiangming Xiao. Students (and now colleagues) with whom I have worked on disease questions include Carlos Eduardo Almeida, Sair Arboleda, Lindsay Campbell, Chris Ellis, John Giles, Carlos Ibarra-Cerdeña, Ryan Lash, Sean Maher, Yoshinori Nakazawa, Simon Neerinckx, Abdallah Samy, and Richard Williams, although surely I learned much more from them than they from me about these topics. Lance Waller provided crucial and illuminating discussions of many ideas handled in this book. Finally, the numerous members of the University of Kansas Ecological Niche Modeling Group provided frequent insights and ideas regarding methodological challenges. Once again, any omissions are ones of accident and not purpose.

I thank my lifetime friends and colleagues at the Centro de Referência em Informação Ambiental (CRIA), in Campinas, Brazil, for hosting me during the preparation of the bulk of the manuscript for this book. The Institute of International Education's Fulbright Scholar Program generously provided funding for that research stay. Rosa M. Hernández supplied valuable logistical support; Laura Russell helped generously with large-scale online database queries, and Lindsay Campbell assisted with the preparation of worked-out examples and figures.

A big thank you also goes to Vincent Burke and Catherine Goldstead of Johns Hopkins University Press, for their expert assistance in the design, conception, and development of this book, and to Kathleen Capels for detailed and incisive editing of the entire manuscript; the book would not have been possible without their enthusiastic participation. Maps were prepared by Lucidity Information Design, LLC.

My family—Rosi, Mily, Danny, Cassidy, Khaleesi, and Danerys— showed enormous patience with the long absences involved in preparing the manuscript for this book. They know the appreciation that I have for them, but their support, interest, and enthusiasm were and always will be fundamental to any and all of my achievements.

Finally, and far from least, a big thank you goes to my sister, Ellen Salehi. For the earliest encouragement that she gave me toward science in general, and disease biology in particular, I owe her an enormous debt of gratitude, which I offer in the form of this acknowledgment.

1

Introduction

This volume treads a narrow path along the border between several worlds. I am a biodiversity scientist by training, and my work with diseases has been possible thanks only to the counsel of others who are much wiser about this field. Epidemiologists will be tempted to dismiss this book as "out of left field," while ecologists and biogeographers may dismiss it as overly applied. In actuality, though, the volume takes realistic goals—the development of predictive maps allowing an anticipation of disease transmission risk—plus a conceptual foundation in ecology and biogeography, and sets out to derive new methodologies that respond optimally to the challenge. The hope is for synthesis, and benefits for each of the worlds that this book touches.

Ecology and Biogeography

This book sets out a conceptual and empirical framework for a rather serious shift in how disease transmission risk is mapped in the fields of public health and epidemiology. For the first time, disease transmission risk is looked at as what it really is: the distribution of a species (the pathogen) in space and time in a complex abiotic and biotic context—that is, the environment in the broadest sense (Peterson 2008a). Treatments to date have focused in largest part on the spatial dimensions of the question, so that transmission risk is often handled as a purely geographic phenomenon.

I argue that this path is perilous: species' distributions are multidimensional phenomena, and spatial dimensions are—if anything—a more indirect construct of the causal factors, which are environmental and biotic in nature.

Two branches of biodiversity science have much to offer to this challenge of mapping the transmission risk of diseases: ecology and biogeography. Ecology has long focused on environmental conditions as important determinants of distributions of species (Hutchinson 1957). The suite of conditions under which a species can maintain populations can be referred to as its ecological niche (more precise definitions will be provided in chapter 2), although this term has been used in diverse ways, which has clouded its utility (Chase and Leibold 2003; Peterson et al. 2011). Nonetheless, the niche concept has served as an excellent basis for thinking about why a species can maintain populations where it does.

Biogeography is a field that had a very different set of origins. Pre-evolutionary thinking had trouble with the idea that all species were not distributed everywhere, such that Alfred Russel Wallace's detailed distributional observations (Wallace 1860) proved controversial (Smith and Beccaloni 2008). Quite soon afterward, however, Victorian-age natural historians and systematists explored unique regional faunas with immense energy, and a broad documentation of global biodiversity emerged, partitioned into many regional faunas that overlapped either only partially or not at all (e.g., Godman and Salvin 1879–1915). Development in biogeography progressed rapidly thereafter, with the introduction of ideas from the New Synthesis (Mayr 1942), historical biogeography (Nelson 1974), island biogeography (MacArthur and Wilson 1967), and phylogeography (Avise 2000), to name a few major advances.

Ecology and biogeography have nonetheless remained uncomfortable bedfellows. Although the niche concept was created by a systematist, Joseph Grinnell, in papers that were much more biogeographic than ecological (Grinnell 1914, 1917, 1924), the two areas diverged rapidly; little common ground was found until relatively recently (e.g., Peterson et al. 1999; Warren et al. 2008; Evans et al. 2009). Indeed, the two fields have begun to relink, with initial papers pointing out the need for considerations from both to explain species' distributions (Pulliam 2000; Soberón and Peterson 2005), and subsequent contributions providing much more detail. Although differences still persist, the emerging field of ecological niche modeling serves to unite and integrate the two broad approaches: species' distributions are seen as the joint result of ecological/environmental limitations (niches), geographic constraints, and historical contexts.

This Book

The present volume brings the spatial epidemiological exercise of mapping disease transmission risk into the context of this ecology-biogeography interface. A disease transmission system is seen as a set of species that interact in hosting, vectoring, and moving a pathogen among different biological realms—one of which turns out to be a human or a human-associated animal. Each of these participating species has its own biogeography—linkages between its ecological niche (conditions appropriate for its survival and reproduction), its interactions with other species, and the area that has been accessible to that species over relevant time periods. All of this complexity is played out over the highly diverse, multiscalar topography of real-world landscapes.

Under these rubrics, "transmission risk" is seen as the combination of presence of the pathogen with the requisite factors that lead to infection of a human or other animal of concern. A large part (though not the only part) of transmission risk is simply the intersection of the distributional areas of the species involved, so much of this book will concentrate on creating an appropriate conceptual foundation and empirical plan for mapping distributions of species that are relevant to diseases. Other factors (e.g., human population, human socioeconomic status, population immunity, etc.) are considered to be modifiers of the basic distribution of the pathogen leading to disease transmission. The scope of this volume includes both human and animal diseases, since the processes involved are the same, particularly under the ecology-biogeography framework. I emphasize, however, that this book is not intended as a comprehensive literature review (either of spatial epidemiology or of ecological niche modeling). Rather, I intend to present a synthesis of a novel viewpoint and methodology in mapping disease transmission risk in spatial epidemiology.

The traditionalists in spatial epidemiology might balk at many points in this volume. They may worry about abandoning approaches that have long been employed—the distinction between potential and actual distributions and the spatial-to-environmental shift will be particularly difficult challenges in this regard. Comments have already been made along these lines (e.g., Coetzee 2004), which is only to be expected with any drastic methodological shift. My impression is that when traditionalists take the time to listen, and when the concepts and approaches are explained carefully and rigorously, misunderstandings can be avoided and a new, synthetic, and novel understanding can be achieved. The ecological and biogeographic approaches outlined in this book offer considerable advantages in terms of the detail that is possible in mapping, as well as in terms of its predictive ability—these two features in and of themselves should make reading this book worthwhile to those in the field. For those who remain unconvinced, I simply ask for patience and an open mind in reading what I discuss here.

Conclusions

Many major steps have already been made, thanks to forward progress from the biodiversity world. Scientists in this field have been assembling, digitizing, enabling, and sharing key biodiversity data sets for decades, and this enterprise has grown massively in recent years. In parallel, these scientists and ecologists have been exploring and improving tools for understanding and anticipating geographic and environmental distributions of species. These techniques are generally termed ecological niche modeling, and they offer an exciting new possibility for broadly based views of species' distributional potential. These advances, and the beginnings of their uptake in epidemiology and disease biology, constitute the first steps in the direction of the synthesis offered in this book.

Finally, as will become clear in the course of reading the volume, the job is far from complete. Much remains to be understood and clarified regarding critical steps in the reasoning that underlies these methodologies. How does abundance relate to niche and distribution? When, why, and how do species interact? What does disease "emergence" mean in the context of ecology and biogeography? These questions are key, yet they are only beginning to be explored in disease-related mapping efforts. The ideas presented in this book, then, are a work in progress, and much remains to be learned.

My first applications of ecological and biogeographic thinking to diseases were published more than 10 years ago (Beard et al. 2002; J. Costa et al. 2002; Peterson et al. 2002b), yet genuine synthesis still seems to lie down the road. Many disease researchers, public health researchers, and epidemiologists are beginning to look at and explore new methodologies for mapping disease transmission risk, such as the niche-modeling approaches explored here, yet they do not always have the framework provided by experience in ecology and biogeography. This book aims to develop a firm conceptual basis and an appropriate corresponding methodological synthesis for such explorations. My hope is that equipping many additional researchers in the field with such a platform (ecology + biogeography) will focus better minds on the challenge, and lead to better solutions even more quickly.

PART I

DISTRIBUTIONAL ECOLOGY

2

General Conceptual Framework for Species' Distributions

This chapter aims to provide a general conceptual framework for thinking about geographic distributions and the ecology of species. In particular, it presents a summary of general suites of factors that influence why species occur or do not occur at specific sites. To be able to think effectively about these questions, it is necessary to have a basic understanding of what an ecological niche is, as well as to appreciate the linked spaces (environmental and geographic) in which such questions must be pondered. A three-part framework of biotic, abiotic, and mobility (BAM) factors can guide thinking about where species are and are not (and why); examples and a discussion of their importance are provided.

Historical Background

The numerous factors that determine areas of distribution of species have been the subject of decades-long debates and misunderstandings in the fields of ecology and biogeography. Grinnell (1917, 1924), in introducing the niche concept to the biological world, presented lists of factors that played potential roles in the process, focusing principally on abiotic features of the environment, although he did mix in a few biotic dimensions, such as the availability of key food items. He also appreciated the role of geographic barriers in shaping species' distributions (Grinnell 1914), a point that subsequently got lost for a number of decades.

The niche concept proved to be such an attractive suite of ideas that it got co-opted for other purposes in the then-budding field of ecology. Grinnell's (1917, 1924) original term for a concept that was focused on species' requirements in terms of abiotic conditions was adopted to apply more broadly to a species' role in a biotic community (Elton 1927). These dual meanings for a single term, referring to two separate (but both very important) concepts, have continued in the literature up to the present, and confusion still reigns as to which concept is being used when one says "niche" in ecology (compare, e.g., Chase and Leibold 2003; Soberón 2007). This confusion was consummated in a paper published with the title "I can't define the niche but I know it when I see it" (Godsoe 2010), which illustrates the situation quite aptly.

Returning to the historical sequence, Hutchinson (1957) generalized the idea of niche to refer to a species' distribution in a multidimensional space that is defined in environmental dimensions. Some of these dimensions are abiotic in nature, whereas others summarize interactions with other species; Hutchinson (1978) referred to these two types of environmental dimensions as "scenopoetic" and "bionomic," respectively. Hutchinson (1957) conceptualized the distribution of a species as the geographic reflection of its fundamental ecological niche (the niche with respect to scenopoetic variables), reduced by bionomic variables (interactions with other species), to become the realized ecological niche. (Note that, formally put, the terms "abiotic" and "biotic" turn out not to be completely appropriate; a better statement is noninteractive versus interactive variables, but this distinction is not emphasized here [Peterson et al. 2011].)

Since Grinnell's early work, what was not emphasized for several decades in the field of ecology was the role of geography, in particular geographic barriers, in shaping species' distributions. Although biogeographers were highly focused on the part barriers played in the range-splitting process they term "vicariance" (Nelson 1974; Wiley 1981), ecologists in general did not take these spatial factors into consideration. It was not until papers by Pulliam (2000) and Soberón and Peterson (2005) that these factors were explicitly brought back into distributional ecology. Grinnellian niches have seen a renaissance in the massive recent use of tools for what is called "ecological niche modeling"—correlative estimates of environmental conditions under which species are able to maintain populations—that focus along the lines of niches as requirements and form the basis for much of this book. In the end, then, a more modern conceptualization of the distributional ecology of species would include consideration of biotic factors, abiotic factors, and mobility (BAM), joined together in a synthetic view I refer to as the "BAM framework" (Peterson et al. 2011).

A General Schema of Distributional Ecology

In addition to the historical introduction above, what is necessary to this book is a careful and comprehensive conceptual framework for thinking about and discussing ecological niches and the geographic distributions of species. I use explicit and formal mathematical terminology where prudent (but avoid excesses as much as possible) to define particular niches and their spatial manifestations as geographic distributions. For a more complete treatment of these topics, I refer the reader to a more general presentation of relationships between ecological niches and geographic distributions that I published recently with a group of coauthors (Peterson et al. 2011).

What Is a Fundamental Niche?

The fundamental ecological niche of an organism, referred to as $\mathbf{N}_F$ in this book, is a reflection of physiological limitations. Organisms have sets of tolerance limits and requirements with respect to scenopoetic variables (e.g., temperature, growing days, humidity). The field of physiological ecology focuses on understanding and characterizing organisms' responses to external conditions.

Generalities that have emerged from multitudes of physiological ecological studies include the following. First, most such studies have found that physiological response curves tend to be bell shaped. When organismal responses are characterized over a broad-enough range of values, extreme values at either end of the curve tend to be unsuitable for those organisms, and a unimodal hump is observed at intermediate values (Birch 1953; Green 1971; Maguire 1973). Some of the unsuitable extreme values can be rather absurd—aquatic mosquito larvae may not survive well in either ice or boiling water—but the important point is that these tolerance curves are generally hump shaped.

A second point is that, in spite of the rather regular shape of most physiological response curves, niches can be quite diverse. The tolerable interval for an organism can be broad or narrow, making the species effectively a generalist or a specialist with respect to different environmental dimensions. Niches, extending from single-variable to multivariate tolerances (effectively Hutchinsonian niches), should take the form of convex, ellipsoid, multivariate clouds, which may be highly multidimensional (a spheroid) or simpler and unidimensional (an elongate shape). In other words, fundamental ecological niches may assume a great variety of shapes, and this diversity will be augmented still more when conditions that actually exist on real-world landscapes are considered (see below).

Finally, physiological tolerance curves are an inherited characteristic of organisms and have evolved as a result of natural selection. Many studies in evolutionary biology have documented the heritability, the process of natural selection, and the adaptive evolution of organisms' environmental tolerances (Huey and Kingsolver 1993; Lee et al. 2003; Kellermann et al. 2009). Perhaps the clearest illustration of the genetic basis for these traits are the many examples from agronomy, in which artificial selection on key tolerance traits has successfully broadened the sets of environments under which many crop plants have been able to survive and reproduce (Bolaños et al. 1993; Ashraf 2004).

Geographic-Environmental Space Duality

Species are distributed on real-world landscapes that, in effect, serve as the stage on which niches are manifested. In this sense, one can ponder a bidirectional interaction between scenopoetic environmental spaces (where niches exist) and geographic spaces (where geographic distributions exist). At any point on the surface of the Earth, one can characterize the environments manifested there: climate, soil characteristics, solar radiation, and the like. The conditions and the places are linked via their spatial positions.

So we have two spaces—geographic and environmental—that are joined together, because every point on Earth occurs in both; these two spaces will be referred to as **G** and **E**, respectively. It is worth pointing out that, in relatively simple environmental characterizations, a single environmental combination (a single point in **E**) may correspond to various sites in **G**; as environments are characterized in more and more detail, however, the correspondence between the two spaces becomes one-to-one, and each point in **E** maps uniquely onto a single site in **G**. This space-to-space mapping process (tracking a point or a region in one space as it corresponds to points or regions in the other) can be formalized as follows: the environments corresponding to a given point x in **G** are termed $\eta(x)$, while the site(s) in **G** corresponding to a given environment y in **E** are termed $\eta^{-1}(y)$.

Hence ecological niche models plot known occurrences of species in geographic space, and then infer some form of distribution in environmental space. Sectors of **E** thus identified can then be taken back to **G** to delimit potential distributional areas. Various algorithms are used to fit these models; all characterize in some way the part of **E** that is within or close to this cloud of known occurrences in environmental dimensions (Peterson et al. 2011). In the most basic sense, one could take the multivariate ellipsoid nature of fundamental niches and simply find the ellipsoid that includes all (or most) of the known occurrences; such an ellipsoid could be considered a hypothesis of the fundamental niche. Nonetheless, the real world complicates things considerably, as will be appreciated below.

The BAM Framework: Different Niches and Distributions

The historical background presented above touched on many of the critical points for this more synthetic view of species' distributions. Here, a useful heuristic tool, the BAM diagram, will be explored in greater depth. A particular focus will be on naming different niches and distributional areas and understanding how they relate to one another. Recall that Hutchinson (1957) conceived of species' distributions as determined by their fundamental ecological niches, and reduced by biotic interactions, to produce realized ecological niches. Here, I formalize this description and make it explicit in terms of the linked geographic and environmental spaces, but I also add mobility considerations to the effects of landscapes and geographic spaces.

In geographic space, we can begin by defining a series of areas. The overall region under consideration is **G**. Some portion of that area presents suitable conditions for the species of interest, here termed **A**; that is, **A** represents the set of sites x for which $\eta(x) \subseteq \mathbf{N}_F$. Thus, in some sense, **A** represents the full distributional potential of the species within **G**.

Yet species frequently do not inhabit the entire spatial footprint of their potential—that is, their ranges cover less than the full extent of **A**—for a multitude of reasons (Peterson 2003). One extremely important factor is that of mobility: has the species in question had the opportunity to send dispersers to the site, such that it would be inhabited if it were suitable? This mobility area, termed **M**, will frequently be much larger than the actual distribution of the species, as many species send out dispersers to test the waters across broad areas that are not necessarily suitable for the species but that are close to occupied distributional areas (Pulliam 1988). A detailed treatment of **M** and its pervasive but cryptic effects on niche-modeling applications has been presented in Barve et al. (2011).

Biotic interactions may also play a role in shaping distributional areas. The species in question may have particular requirements in terms of other species: the presence of specific food items, reservoirs, or vectors; the absence of certain predators, pathogens, or parasites; the presence of some specific biotic substrate; and so forth. The area where these requirements are met can be termed **B**. (More discussion about the constraining role of **B** is presented below.)

Now we are ready to define some distributional areas. Within **G**, the portion that is habitable for the species in the absence of considerations of other species is **A** (sometimes presented as $\mathbf{G}_A$). The portion of **A** that is also suitable in biotic terms is the potential distributional area $\mathbf{G}_p = \mathbf{A} \cap \mathbf{B}$. Frequently, however, not all of $\mathbf{G}_p$ is accessible to the species; hence we can define the actual (occupied) distributional area as $\mathbf{G}_O = \mathbf{G}_p \cap \mathbf{M}$. The remainder of $\mathbf{G}_p$, the part that is not accessible to the species, is termed the "invadable distributional area," or $\mathbf{G}_I$.

Turning to niches, a few more definitions are in order. $\mathbf{N}_F$ has been defined above: it is the fundamental ecological niche, and many results from the field of physiological ecology lead us to expect that $\mathbf{N}_F$ will form a convex, and rather regularly shaped, cloud in environmental space and will be hump shaped with respect to any single environmental dimension (Birch 1953; Green 1971; Maguire 1973). In terms of physiological responses, this simple and consistent response shape is quite logical—for most or all requirements, species will require more than the minimum and less than the maximum that could possibly exist, and scenarios in which multiple optima would exist are difficult to envision.

It is important to bear in mind that representation of the full breadth of environments that make up a fundamental niche within $\mathbf{M}$ for a given species and study area will be rare (Saupe et al. 2012; Owens et al. 2013). Rarely, if ever, will a fundamental niche be able to be characterized fully in correlational studies. Rather, in most cases only a subset of the environments that make up $\mathbf{N}_F$ will actually be represented within the area of analysis; this set of environments determines the portions of the fundamental niche that are actually observable (Owens et al. 2013).

If we make the simple assumption that the study area is equivalent to $\mathbf{M}$, then the portion of $\mathbf{N}_F$ that can be observed in a study and characterized via niche-modeling efforts is $\mathbf{N}_F \cap \eta(\mathbf{M})$—that is, the fundamental ecological niche reduced by the set of environments represented within $\mathbf{M}$, which I will refer to as $\mathbf{N}_F^*$. This set of environments is called the "existing fundamental ecological niche," or that part of the fundamental ecological niche that is actually represented on landscapes that are relevant to the distribution of the species. It is especially important to note that $\mathbf{N}_F^*$ is not necessarily regular or convex or simple in shape; rather, given the vagaries of representation of environments across real-world landscapes, $\mathbf{N}_F^*$ may be exceedingly irregular, and its correspondence to and representation of $\mathbf{N}_F$ may be quite incomplete (Soberón and Peterson 2011).

A final detail is the effects of biotic interactions on these niches. Although it has been argued in biodiversity circles that $\mathbf{B}$ will be manifested chiefly on fine spatial resolutions, and that most coarse-resolution distributions of species will be affected little by biotic interactions (Soberón 2007; Peterson et al. 2011), these arguments may not hold as firmly for disease transmission systems (Peterson 2008a). Here again, the environments under which suitable biotic-interaction conditions $\eta(\mathbf{B})$ occur may reduce the ecological niche still further. We can use Hutchinson's terminology, and refer to this quantity as the "realized ecological niche"; it would be defined formally as $\mathbf{N}_R = \mathbf{N}_F^* \cap \eta(\mathbf{B}) = \mathbf{N}_F \cap \eta(\mathbf{M}) \cap \eta(\mathbf{B})$.

A final point regarding the BAM framework is that the BAM diagram may take on several different configurations (Saupe et al. 2012). The Hutchinsonian view of the world would have everything accessible ($\mathbf{M}$ being large with respect to $\mathbf{A}$ and $\mathbf{B}$), which implies that $\mathbf{G}_P = \mathbf{G}_A$, and that $\mathbf{G}_I$ is nil. A quite different view of the world might have a very limited $\mathbf{M}$, such that $\mathbf{A}$ is not particularly limiting, and the species' geographic distribution is structured mainly by mobility questions, not by suitability considerations. These BAM configurations influence strongly which modeling efforts will have the potential to be successful, putting a premium on thinking carefully about BAM configurations at the outset of a study, since some configurations present impossible situations for the development of effective and predictive models (Saupe et al. 2012). (The "Beale Fallacy" and related issues will be discussed in chapter 11.)

Disease Systems

To bring this rather broad, conceptual discussion back a bit closer to questions of disease transmission, in this section I examine some disease-specific examples and considerations. To begin with, aspects of ecological niches have now been explored for a number of arthropod disease vectors, which provides examples for at least certain elements of some transmission cycles. To give one illustration, figure 1 shows laboratory measurements of temperature-related durations of aquatic stages of the Afrotropical malaria vector mosquito *Anopheles gambiae* (Bayoh and Lindsay 2004). Note that these measurements yield the expected intermediate hump of suitable conditions, but also note that they were developed across the full spectrum of relevant temperatures (the niche saw no reduction because of existing conditions), and that they were developed in laboratory settings, where influences of other species most likely were minimal. Measurements of the realized niche for this species may turn out to much less regular and convex.

Response surfaces for other physiological characteristics, and for other species in other disease transmission systems, have proven to be similar. For some characteristics, such as humidity and water availability, one might expect the hump to be at one extreme—mosquitoes like water, right? Yet it turns out that even these insects have their limits: performance drops off even in mosquitoes under the wettest of circumstances (Bar-Zeev 1960).

In one review (Peterson 2008a), BAM configurations were explored explicitly for disease systems, and four differences from "normal" bio-

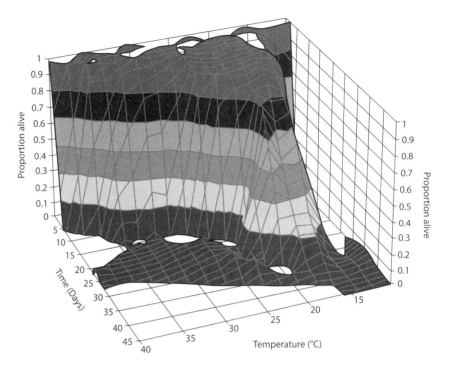

Figure 1. Age-related three-dimensional survivorship curves for the aquatic stages of *Anopheles gambiae* at constant temperature in the range from 10°C to 40°C, illustrating the intermediate optimum characteristic of physiological response curves. From Bayoh and Lindsay 2004, © 2004 John Wiley and Sons.

diversity were pointed out. Given the focus of this book, and in light of the conceptual framework laid out in this chapter, these differences are worth mentioning and discussing a bit, so the reader can reflect on BAM configurations that are of relevance to disease systems of interest. Some examples of BAM configurations are also presented in Saupe et al. (2012).

Interactions rule. In the biodiversity world, it has been argued that only rarely will biotic interactions change the gross aspects of species' distributions (Soberón 2007; Peterson et al. 2011), an idea that has been termed the "Eltonian Noise Hypothesis." Under this hypothesis, species interactions are manifested on very fine spatial resolutions (e.g., millimeters to tens of meters), but not commonly at coarser resolutions (kilometers and broader), such that **B** rarely restricts the distributional possibilities of species.

Disease transmission systems, however, may frequently represent situations in which the Eltonian Noise Hypothesis does not hold. Many pathogens show tight one-on-one or one-on-few host relationships, in which a particular pathogen species is only found in a given host. Perhaps the most famous recent example would be the especially close relationship between

Sin Nombre virus and the mouse *Peromyscus maniculatus*, although relationships are apparently not as tight in some other American hantaviruses (Jonsson et al. 2010). Additional seemingly tight pathogen-to-host relationships have been noted in Lassa fever with *Mastomys* rats (Lecompte et al. 2006), henipaviruses with pteropodid bats (Field et al. 2007), and filoviruses with bats in the genus *Rousettus* (Towner et al. 2009). Other disease systems appear to have less stable or less strict host relationships, and some may show only associations of convenience, in which a host becomes infected when it occurs in the same area as the pathogen. Maher et al. (2010) presented a detailed and quantitative analysis of such a case, with plague-host relationships in western North America.

Stable and unstable interactions. Biotic interactions, in situations where they are important in shaping the geographic distributions of pathogens, can show completely different behaviors in different situations. For instance, West Nile virus was long known as a "minor" virus causing occasional febrile illness in Africa, southern Europe, and the Middle East (Work et al. 1955; Taylor et al. 1956). The virus is hosted by a broad diversity of bird species and had never been seen to cause either extensive avian mortality or widespread human illness (Work et al. 1955). Then, around 1998, surprisingly different pathogen-bird interactions were noted, with wide-ranging mortality in birds in Israel and later in North America (Lanciotti et al. 1999; Peterson et al. 2004c). Similarly contrasting pathogen dynamics have been documented in considerable detail for plague (Gage and Kosoy 2005), and numerous other examples could be added to this list.

A *can be of minor importance.* Many pathogens spend all—or nearly all—of their life history within hosts, which can make coarse-resolution scenopoetic environmental dimensions somewhat irrelevant, at least during parts of the transmission cycle. For endothermic (warm-blooded) hosts, the internal homeostasis of the host may create microenvironments that are perfectly suitable for the pathogen, regardless of the conditions that are manifested outside of the host's body (such as a flu infection in a human being). Under such circumstances, the distribution of a pathogen may have less to do with the scenopoetic conditions being correct directly for the pathogen than with the scenopoetic correlates of **B**, in effect the niche of the host or vector and not of the pathogen per se.

Such situations are simply scale issues. At a fine temporal scale, the pathogen's distribution depends on environmental conditions existing on microscopic scales, such as the acidity of the surface of the intestine. At broader extents, it may be independent of scenopoetic conditions, but, viewed over the entire transmission cycle, scenopoetic correlates of distributions may emerge: the host may itself have some environmental limitations (environmental manifestations of **B**), or some brief transmission stage may be a bottleneck in transmission, such as person-to-person transmission of influenza (Weinstein et al. 2003). These details create what is a scenopoetic ecological niche for the pathogen and its transmission cycle, mediated by the conditions associated with **B**, and make the application of ecological niche-modeling approaches possible. Yet these situations can be quite complex, and they may have nontrivial implications, so the details must be pondered carefully at the outset of a study.

M *can be of variable importance*. Dispersal restrictions, termed **M** in the BAM diagram, can be a highly episodic phenomenon in disease transmission systems, such that disease dynamics may change dramatically over time. Consider the SARS epidemic. SARS had existed in Southeast Asia all along, apparently hosted by *Rhinolophus* bats (Guan et al. 2003; Li et al. 2005). Had public health geographers been studying the SARS coronavirus in 1990, they might have argued that **M** for this virus would prove to be fairly narrow—the virus was probably limited by the movements of its host lineages, and thereby had never extended outside of Southeast Asia. With a host shift (jumping to humans), however, the virus was able to spread globally exceedingly rapidly (Lloyd-Smith et al. 2005) and thus enjoyed an especially broad **M**, at least for a short time. These shifting roles of **M** are quite critical in calibrating models correctly, with hypotheses of **M** essentially predetermining the results of all such modeling exercises (Barve et al. 2011), so careful consideration is again in order.

Conclusions

Mapping disease transmission risk is, at least in part, an exercise in mapping pathogen distributions. GIS programs are now particularly user friendly, and putting dots or shapes on maps is pretty easy, but that does not make the maps useful. Rather, effective risk maps will have some predictive power regarding where disease transmission is likely to occur. These qualities will emerge when risk maps are developed within an appropriate conceptual framework, with the ecology and biogeography of species' geographic distributions considered carefully, so the maps can best estimate the quantity of interest. The BAM framework provides a useful heuristic tool in this regard: species' distributions are conceptualized as the interaction between biotic, abiotic, and mobility (BAM) considerations. Disease transmission systems may be quite atypical of biodiversity more generally, so a thoughtful examination of BAM configurations is key in guiding researchers toward an appropriate design for data gathering, analysis, and interpretation.

3

Status of Data for Understanding Disease Distributions

In the remainder of this book, the reader will come to appreciate the degree to which the analyses that are proposed are data limited. That is, the models themselves are often fairly simple and may not take huge amounts of effort to calibrate, evaluate, and interpret. Assembling and assigning geographic coordinates to the occurrence data that are a critical input to such analyses, on the other hand, can be an enormously time-consuming and difficult task. In this chapter, I give an overview of the status of various data sets that are crucial to efforts to characterize disease distributions and map transmission risk, as well as present some basic implications of these considerations for methodologies in mapping transmission risk. The main point is that a premium should be placed on making primary, research-grade data openly available to the broader scientific community.

Disease Case-Occurrence Data Sets

Large quantities of data documenting disease case occurrences do exist, as disease has been the focus of the medical and public health professions for centuries. Problems remain, however, on at least three levels.

1. *Archival documentation*. Data must be recorded in such a way that information is not lost in the process. Too often, data have been recorded for a given study only, or under idiosyncratic formats that may not

cover all bases. The biodiversity community has solved these problems via careful development of controlled data architectures that communicate critical data fields in standardized formats for any biodiversity record (TDWG 2007a, 2007b).

2. *Integration*. A direct consequence of the lack of a formal structure for disease case-occurrence data is that melding data from different sources can be quite complex, which makes it difficult to build data sets on scales that exceed the efforts of a single research group. Again, the biodiversity community has managed these challenges effectively via the development of protocols by which data stored on different platforms, and even with different field structures, can nonetheless be integrated effectively (Blum et al. 2001).

3. *Opening access*. This aspect of the disease data challenge is perhaps the most difficult of all, in large part owing to privacy concerns. A few initiatives, however, have found a way to make this idea work (see below). Data with reduced levels of detail could be disseminated openly and freely, and more precise data could then be made available on request, with appropriate permissions and documentation of how privacy will be protected. The biodiversity community has developed massive, open-access, integrated (but distributed) data resources: indeed, these resources are the basis of most of the examples given in this chapter. (Note that the examples provided here are based on queries in May 2012—although the numbers will have changed, the general picture will not).

Given that the medical, public health, and epidemiological communities have, in large part, failed at this three-point challenge, the state of disease case-occurrence data is generally dismal. Relatively little of it is available freely and efficiently, which frequently necessitates collaboration with institutions and individuals who have developed relevant data sets. While not a bad thing in and of itself, such proprietary data management makes for little freedom of experimentation and exploration in the field. (Some of the relatively few open-access data sets are reviewed in the paragraphs that follow.)

Among the best of the data resources that are online is the Influenza Research Database (Squires et al. 2012), which (among other things) has a particularly thorough summary of influenza surveillance data that are available to download. An example of avian influenza surveillance data is shown in figure 2, illustrating 337 influenza detections out of 9,392 bird samples tested across Africa. The important point is that primary, research-grade data are openly accessible for downloading and analysis

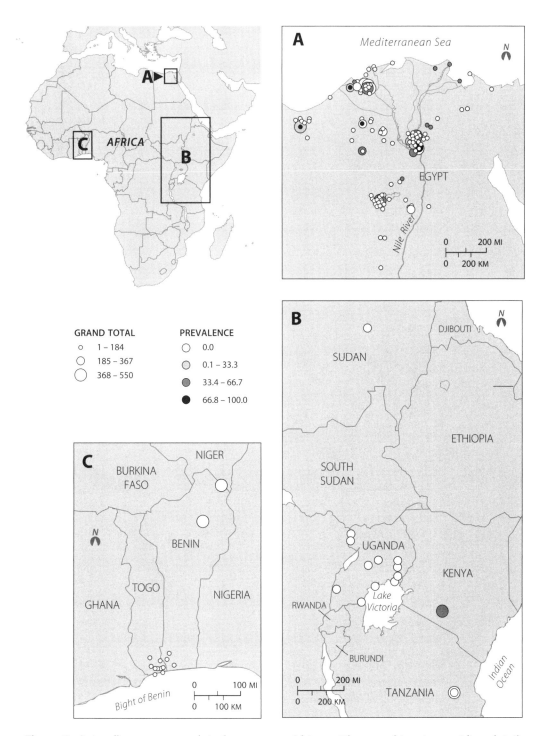

Figure 2. Avian flu occurrence data from across Africa, with several insets providing details for particular regions. The size of the symbols reflects the number of samples tested, and the shading of the symbols shows the prevalence that was discovered. Maps developed from data available from the Influenza Research Database (Squires et al. 2012).

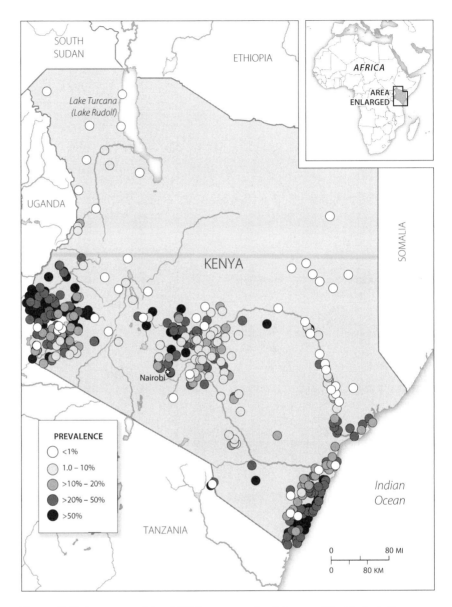

Figure 3. Data summarizing 1,034 surveys of soil-transmitted helminth occurrences across Kenya, 1974–2009. Modified from the Global Atlas of Helminth Infection, www.thiswormyworld.org.

by interested persons. The Mapping Malaria Risk in Africa (MARA) initiative (www.mara.org.za) also provides open access to useful information on malaria incidence (e.g., prevalence rates, vector occurrences), although the site has not been updated recently. Finally, the Global Atlas of Helminth Infections site (www.thiswormyworld.org) provides excellent, map-based summaries of detailed data on helminthic diseases (figure 3); the data underlying these mapped summaries have been made readily available on request.

Those sites are the good ones. Many other data resources for disease case-occurrences consist of secondary, summarized information, which turns out not to be at all useful for modeling efforts. The Food and Agriculture Organization (FAO) has a H5N1 HPAI Global Overview site that aims to provide data on avian influenza occurrences, but the information is in textual format and not at all friendly for extracting data. Similarly, the World Health Organization (WHO) supplies country-level summaries for a number of diseases (http://apps.who.int/gho/data/), but the site gives no access to the underlying data; figure 4 is an example summary from this source for cholera.

Beyond that, at times, data can be created by recycling information from other sources. For example, occurrence data can be drawn from the National Center for Biotechnology Information's (NCBI) GenBank (created as a source for genomic information) or from the International Society for Infectious Diseases' ProMED archives (created for communication about disease outbreaks). Finally, and perhaps most disappointing, the amply funded PREDICT initiative of the United States Agency for International Development, begun in 2011, was established to build "a global early warning system to detect and reduce the impacts of emerging diseases that move between wildlife and people," but project personnel have not yet found a means of putting the data online. The long and the short of the situation is simply that disease case-occurrence data are not abundantly or effectively available, unless one either collects the data oneself or collaborates with people who are gathering the data.

Relevant Biodiversity Occurrence Data Sets

The broader biodiversity world has made significantly more progress in preparing large-scale, primary, research-grade data sets for diverse sorts of analyses. This work began in the early 1990s with the FishGopher project and ramped up dramatically by the late 1990s with the Species Analyst effort (Soberón 1999; Stein and Wieczorek 2004). Now, with massive projects such as the Global Biodiversity Information Facility (GBIF), VertNet, speciesLink, Atlas of Living Australia, and others, literally hundreds of millions of data records are available readily and efficiently online—for free—to any user anywhere in the world. What is more, most of these initiatives use common protocols, so data can be integrated and recycled among multiple initiatives, and each network can take maximum advantage of all of the relevant data records, which are digital and able to be

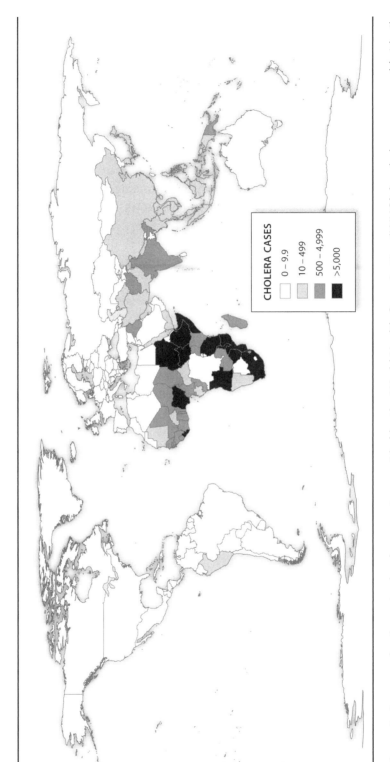

Figure 4. An example of country-level summaries provided by the World Health Organization (WHO) for cholera cases worldwide, illustrating the coarse-grained, polygon-based nature of these data.

shared worldwide. These resources are now sufficiently large to provide a useful starting point for many analyses.

Even such an extensive data resource, however, is not without problems. In particular, the data are dramatically unbalanced in many respects. Certain taxa that are particularly easy to register dominate the data. Consider birds, where almost all specimens have been digitized and have numerous "citizen scientists" contributing millions of data points yearly; sipunculid worms do not benefit from this popularity. In the following sections, I give some illustrations of the richness of information for some taxa and the paucity for others, concentrating on records for groups of taxa that are of interest for disease applications.

Vertebrate Hosts

Vertebrates are the hosts for most zoonotic diseases, so I will focus on the representation of this group in the overall global digital biodiversity data storehouse. Vertebrates have benefited particularly from the initial U.S. National Science Foundation–funded data networks (MaNIS, ORNIS, HerpNet, FishNet2), which have now been joined to form VertNet. Although these numbers are continually increasing, as of May 2012 VertNet linked 72 institutions that had committed data resources to open sharing, for a total of 84.3 million primary biodiversity records. These data are supplemented by data sets linked to other networks that are not part of the VertNet network, but are openly available.

The downside of these data resources is the degree to which they are biased and uneven. Taxonomically, birds are so much easier to observe and identify that they absolutely dominate the data storehouse (figure 5); other groups, such as mammals (perhaps the single most important group of hosts for diseases of humans and other animals), fall far behind the dominant birds. Spatially, major biases are also present. The greatest proportion of the data comes from North America and Europe; outside of these two regions, data are much sparser (figure 6).

These biases are universal features of biodiversity information, since they reflect the general availability of knowledge, but they pose significant constraints on the uses to which biodiversity data can be put (Yesson et al. 2007). Much work remains to be done to build and improve this resource, and the clearest path to filling gaps (to the extent possible) is to complete the capture of data from the world's largest museum-based scientific collections. For example, only 7,648 records of mammals are available from Sudan in GBIF, yet large numbers of mammal specimens from that country exist in the collections of the British Museum that are not yet digitized, or for which digital records have not been shared with the global

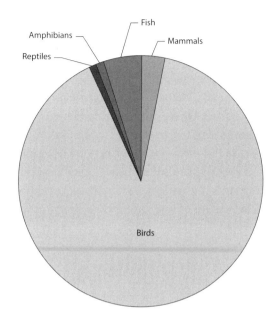

Figure 5. Pie graph of the number of records of vertebrate classes in data mobilized by the Global Biodiversity Information Facility as of May 2012, showing the massive dominance of birds in this data resource.

community. These (and many similar) existing collections have rich data associated with them that simply need to be captured and digitized in an efficient manner.

Arthropod Vectors

The relatively optimistic picture of abundant biodiversity data disappears rapidly when one ventures beyond the well-known vertebrates (and perhaps plants and a few groups of insects). Searching in the largest global data resource (GBIF) and the only resource specific to disease vector taxa (VectorMap), the numbers are nothing short of depressing, with no rich data stores for any taxon. This situation is generally characteristic of data for insects (figure 7), reflecting no observational data and the extreme difficulty of retrospective data capture on huge numbers of specimens with tiny data tags (Lampe and Striebing 2005; Ariño 2010).

The best-documented vector taxon is mosquitoes, with 241,429 records; these records, however, correspond to over 3,500 species and are subject to the same spatial biases in distribution as the broader biodiversity data storehouse. In this case, the biases are not so much toward sites with lots of affluent aficionados, but rather toward areas that have seen intense study and collection *and* from which the existing data have been digitized, integrated, and made available globally. Of course, many more data resources exist. In the United States, for example, the state of Iowa has a long-term monitoring program, from which the data are not only well documented (Sucaet et al. 2008), but are also available online. An-

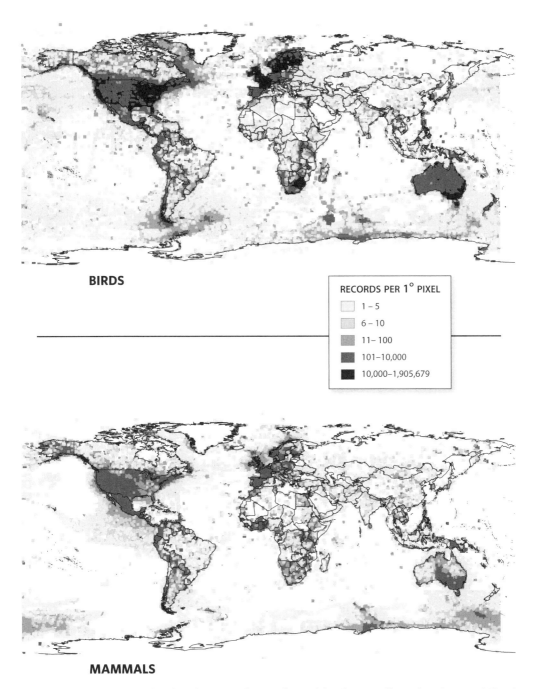

BIRDS

RECORDS PER 1° PIXEL

1 – 5
6 – 10
11– 100
101–10,000
10,000–1,905,679

MAMMALS

Figure 6. The density of bird and mammal records worldwide, as reflected in data mobilized by the Global Biodiversity Information Facility as of May 2012.

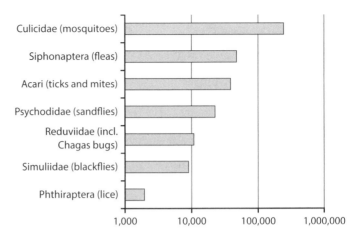

Figure 7. A summary of the representation of different groups of arthropod vectors of disease among the biodiversity data mobilized by the Global Biodiversity Information Facility and MosquitoMap. Note the logarithmic scale on the horizontal axis.

other notable example is the massive database compilation of records of the *Anopheles gambiae* complex across Africa, compiled by Coetzee et al. (2000) and made openly available over the Internet.

A major challenge is the integration of many disparate and nonintegrated vector-occurrence data sets into an overall, much more useful whole. In the GBIF data storehouse, for example, only 22,506 data records are available globally for the hundreds of species of sandflies (the dipteran family Psychodidae). Referring to data available via the speciesLink project, however, 10,614 georeferenced records of sandflies from Brazil alone from the collections of the Coleção de Flebotomíneos in the Fundação Oswaldo Cruz are readily available (figure 8). The complementarity between the two data sets is clear, but the challenge is to search out each of the relevant data sets, work out the access parameters, and integrate the data, which is not a trivial process (for a parallel avian experience, see Navarro-Sigüenza et al. 2003).

Pathogens

The data-searching process becomes still more depressing when one attempts to find rich data records for pathogens. Biodiversity-oriented data resources are focused on biodiversity, generally from the standpoints of conservation and systematics. As a result, entering data on microorganisms has not been a high priority, and they are not well represented in what data are available.

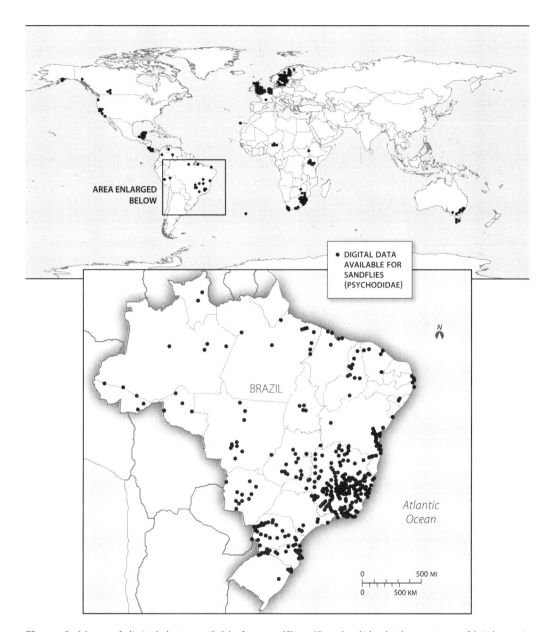

Figure 8. Maps of digital data available for sandflies (Psychodidae), the vectors of leishmaniasis. The global map shows the distribution of data from the Global Biodiversity Information Facility, while the inset shows the distribution of data from one collection in the Fundação Oswaldo Cruz in Brazil, demonstrating that regional sources have much to offer in improving the overall availability of data.

Among the GBIF-enabled data resources, pathogens are miserably represented. A total of 652 virus records are included (most of which are not disease related); of the major bacterial and fungal groups, the best represented are the Actinomycetales and the Clostridia (each with more than 20,000 records). Data for these and many bacterial groups, however, are dominated by marine records, where environmental and ecological genomic studies have produced somewhat greater densities of information (Riesenfeld et al. 2004). Not only are these data still sparse in relation to likely diversities in the species being sampled, but the marine taxa are also rarely relevant to disease-related questions.

An illustration of the potential that is out there if one takes the time to discover, assemble, and integrate the data, however, is an impressive and growing collaboration between the speciesLink network in Brazil and Brazil's Fundação Oswaldo Cruz. The former is a leading biodiversity information network, and the latter is a public health research institute that maintains important microbial collections. As the two institutions have built mutual understanding and sought shared goals, an impressive beginning toward a data infrastructure relevant to pathogen distributions has emerged, as can be seen in the relatively rich data sets that are becoming openly available to the scientific community (figure 9).

Georeferencing

Generalities. A prerequisite for any effort oriented at mapping a particular phenomenon is anchoring it to a point on the surface of the Earth. Such geographic references can be textual in nature (e.g., "25 km northeast of Mexico City"), or they can be presented in various numerical formats that generally include x and y dimensions. In the latter instance, any point on Earth can be described in two or three dimensions as latitude and longitude coordinates, sometimes supplemented by information on depth (for aquatic or subterranean systems) or altitude (for aerial systems). At first glance, once a data set has numerical georeferences attached, one would think that one is ready to start an analysis, but the challenges are only beginning. Frequently, data relevant to biodiversity and disease systems are not presented in these convenient numerical formats.

Thorough georeferencing—which can stay with the data record in the long term as permanent documentation of what is known (and not known) about its spatial information—has a deeper and broader set of goals; for that reason, a pair of coordinates is not enough. Rather, georeferencing must (1) represent the location of the site with as much precision (significant digits in the coordinates) and documentation (e.g., a geographic datum) as possible; (2) include information regarding the

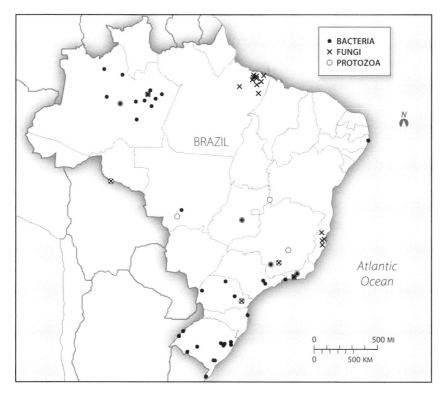

Figure 9. A map of data available from the collections of FIOCRUZ and enabled by the speciesLink network for three groups of pathogens.

degree of knowledge about that precision (or its converse, uncertainty); (3) document the sources of information and the methods used in any and all calculations; and (4) retain all original data, to avoid the possibility of loss of critical information.

Disease occurrence data. Disease data are frequently collected and reported at the level of occurrences in counties or equivalent districts. For instance, reportable diseases in the United States are reported as totals within counties by local and state health officials (L. Eisen and Eisen 2007; Hall-Baker et al. 2011). This approach is certainly convenient, as public health workers at the county level need to provide only simple counts, rather than any more detailed information. Indeed, as will be seen in chapter 4, many current disease-mapping methodologies are designed explicitly for such polygon-based data.

This convenience system has some significant drawbacks, however. In particular, it blinds any analysis of transmission risk to details that are manifested at finer spatial scales than the grain of the polygons. The more precise features of the distribution of the disease are lost because the occurrence data are overly coarse. For example, R. Eisen et al. (2006) analyzed Lyme disease risk patterns across California. They found that analyses at the county level reconstructed the major features of risk across the state, but that significant foci that happened to have small spatial footprints were lost entirely from the analysis (figure 10). A similar loss of detail was noted in analyses of county-resolution data on tularemia occurrences across North America (Nakazawa et al. 2007, 2010).

To offer another illustration of this point, Camargo-Neves et al. (2002) used occurrence data resolved spatially to the level of *municipios* (counties) in São Paulo state, Brazil, to map geographic distributions of *Lutzomyia* sandflies across that particular region. While they had good success in making models that predicted independent data sets better than random expectations, their model predictions had higher levels of omission error than is generally the case in such tests. They attributed this elevated omission error to the broader uncertainty in the occurrence data on which the models were based.

Debate and a solution. Grubesic and Matisziw (2006) noted several problems caused by referencing disease occurrences to ZIP codes in New York State, chiefly revolving around the uncertain spatial footprints that ZIP code districts have, which confuse analyses that depend on area estimates. L. Eisen and Eisen (2007), writing about county-referenced data, made the point much more generally, noting the broad, continent-wide trends that confound such data: counties in the eastern United States are quite small, while counties in the western part of the nation are generally much larger. They argued, correctly, that such broad variations in county size and such coarse overall spatial resolution (to be discussed in more detail in chapter 9) cause serious problems and confusion in analyses of those data.

While L. Eisen and Eisen (2007) got the problem right, they got the answer wrong. Their recommendations were that disease case-occurrence data should be referenced to ZIP codes or census districts instead of to counties, and that critical minimal requirements would be an assessment of whether infection most likely occurred in the peridomestic environment, outside the peridomestic environment but within the county of residence, or outside the county of residence. These recommendations address the questions of spatial resolution and uncertainty only partly and are not sufficient to avoid significant information loss.

In a commentary on the L. Eisen and Eisen (2007) paper, I pointed out two failings in their solution (Peterson 2008b). First, ZIP codes and census districts suffer from the same east-to-west size gradient problems, as they generally reflect human population densities (highest in the east, lower in

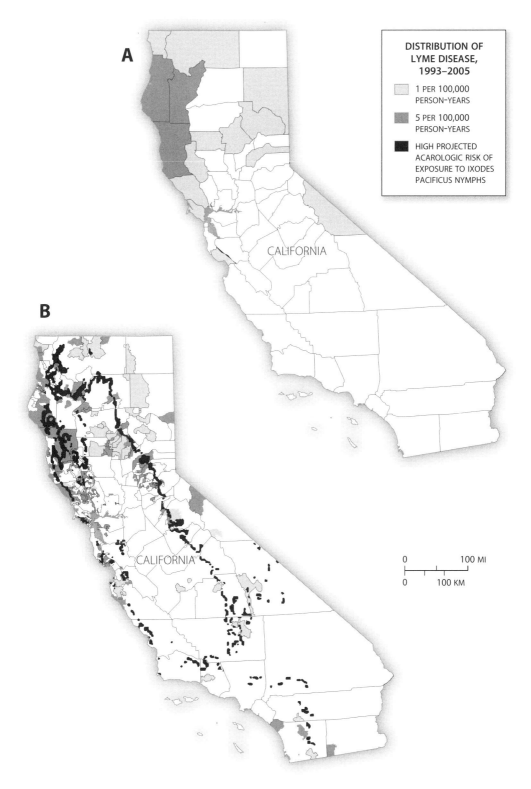

Figure 10. The distribution of Lyme disease in California counties (A) and zip code areas (B) with Lyme disease incidences, 1993–2005. From R. Eisen et al. 2006.

the interior west). Hence moving to finer polygon units for georeferencing does not solve the problem. Rather, it just pushes it back to a somewhat finer spatial resolution, while the same difficulties with coarse and variable spatial resolution remain (Grubesic and Matisziw 2006). The Eisens' suggestion just kicks the can down the road without solving the root problem.

The second failing is in expressing uncertainty. A locality that represents a best guess as to the exposure site may nonetheless be very precise or very vague. Consider a GPS reading at the site where a rabid dog bit a person, contrasted with the general area in which a rancher might have encountered anthrax spores: the two events can be characterized with latitude-longitude coordinates, but they differ markedly in the uncertainty that is associated with the geographic record. More specifically, we can imagine two persons who are next-door neighbors and are both infected with some disease. One might be elderly and housebound, while the other might be a traveling salesperson who moves on a far broader spatial scale. Simply linking those addresses to county names (or ZIP codes or census districts or geographical coordinates) does not distinguish between these two persons, who represent wide contrasts in the associated uncertainty regarding where they were infected. It is worthy of note that this additional, rich information is generally readily available at the time of diagnosis (when the patient is present), but disappears rapidly thereafter.

The field of biodiversity science has put quite a bit of thought and debate into the question of how best to represent biological occurrences and has developed a much more effective solution. Wieczorek et al. (2004) presented a detailed methodology, termed the "point-radius method," in which one's best guess at the coordinates of the actual site are used as the point, and uncertainty is summarized as the radius of a circle that includes the areas within which that point could actually fall. This radius summarizes the degree to which the point is or is not precise, and it can be incorporated into many different analyses (e.g., Peterson et al. 2006a; Nakazawa et al. 2007, 2010). What is more, the point-radius method has been implemented broadly and applied to the challenge of georeferencing the great majority of records that are enabled by the VertNet initiative. Automated software has now been implemented that greatly facilitates the georeferencing process (Guralnick et al. 2006; Rios and Bart 2008; Guralnick and Hill 2009).

The possible objections that could be raised to broad implementation of the point-radius method for disease-related data are (1) the difficulty of educating medical and public health personnel to capture the relevant data; and (2) privacy issues, as the point will frequently be sufficiently finely resolved spatially to identify the person involved (L. Eisen and Eisen

2008). I do not see either as a deal breaker, however. The first problem can be solved via online reporting forms that take advantage of technologies such as Google Earth and Google Maps to allow medical and public health personnel to either enter data in forms or simply draw the data on maps. This first challenge is merely a matter of setting some able programmers to the task and challenging them to create forms or maps where medical personal can efficiently record what information is known or inferred at the time of diagnosis and treatment. The privacy considerations are even easier. For any public reporting, a precise point can be turned back into a county-level reference (or any other higher-order region) and only reported as such. Better data would still exist, and could later be used when privacy considerations are dealt with appropriately through customary permissions. An ideal route by which these ideas and improvements could be explored would be implementation for some test region or regions—perhaps a state or a small country.

The Meaning of No Records

The above paragraphs and pages should serve to convince the reader that presence data, and access to such data, can become significant bottlenecks in the process of developing any geographic understanding of disease systems. One ends up spending more time assembling the data than in actually doing the analyses. In this section, I explore some further implications of these constraints with regard to how one goes about developing those analyses, while taking data complications into account.

We have already seen in the BAM diagram that a species can be present at a site only if the site is suitable in abiotic and biotic terms, and if the site is within the mobility potential of the species in question. (This statement is a bit of a simplification, in that sink populations, misidentifications, and incorrect georeferences are neglected, but those situations should be uncommon.) Put another way, a presence data point can exist only if it falls in the geographic area delimited by $\mathbf{A} \cap \mathbf{B} \cap \mathbf{M}$.

An absence data point, on the other hand, can be the result of several conditions. Very commonly, a species meets the ecological criteria ($\mathbf{A} \cap \mathbf{B}$), but fails the biogeographic criterion ($\mathbf{M}$). This circumstance is what is termed "distributional disequilibrium," in which the species does not inhabit the entire spatial footprint of its habitable area (Peterson 2003). In such situations, areas will exist that will present completely suitable conditions, but they will not be inhabited by the species because of mobility constraints.

Figure 11 presents a probability tree that summarizes the above considerations (which are essentially a restatement of the BAM diagram), as well

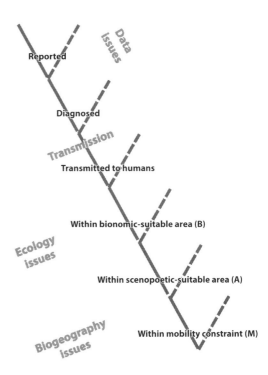

Figure 11. A probability tree summarizing reasons why a given site might be classed as disease-positive for a given pathogen. Six steps are shown (and more could be added easily): the first (*bottom*) is essentially a question of biogeography, the second and third are questions of ecology, the fourth is the issue of whether human infections occur, and the fifth and sixth are issues of information transfer and data handling. At each step, the bifurcation can be seen as a question: "yes" is shown as a solid line, and "no" as a dashed line.

as several other factors that are more disease specific in nature. For instance, a pathogen may be present and circulating in an ecosystem, but it might not be transmitted to humans or their associated animals (Peterson 2008a). Further steps are that a pathogen may be transmitted to human beings, but the disease might not be diagnosed by the available medical facilities, or might be diagnosed correctly but not reported to the broader community.

The upshot of the long probability tree in figure 11 is that many stars must align in order for a disease presence record (or, in a simpler sense, a biodiversity record) to indicate presence at a particular site. Depending on the quality of the presence records, which will be taken into account in the methodologies discussed in chapter 12 as the *E* parameter (Peterson et al. 2008), a presence record must then be accorded considerable weight in the design of our analyses. An absence of records, on the other hand, depends on a multitude of factors and can result from causes that have little to do with the conditions present at a site.

Finally, a model's failure to predict a presence point—called an "omission error"—must then be weighted heavily in our analysis of whether a model is good. A model's failure to predict absence where no presence point is available to document presence (termed a "commission error") is less serious. Filling these knowledge gaps is one major reason why we do these analyses in the first place: a site with no known occurrence of a disease may actually present some risk of transmission. Establishing absence, or at least a high probability of absence, is a complex procedure that must take into account the sampling that underlies a given set of estimates (Anderson 2003). More generally, as the reader will see in chapter 12, the appropriate weighting of omission versus commission errors turns out to be a critical step in improving disease transmission risk maps.

Conclusions

One of the biggest challenges in mapping disease transmission risk is that of assembling sufficient and appropriate occurrence data to document what is known. In some cases, such as with filoviruses, which are extremely rare and poorly known (Peterson et al. 2006a), these difficulties result from the rarity of the transmission phenomenon itself. In other cases, data have been collected, and probably in sufficient quantities, but the disorganized, fragmented, and noncooperative data culture in the public health world prevents any easy assembly of relevant data sets. It is not an understatement to say that the availability of occurrence data constitutes the single most significant bottleneck in the processes and methodologies recommended in this book. More generally, presence data merit more careful consideration than an absence of data, a point that will be highly relevant when considering which tools to use (and how to use them) in producing transmission risk maps.

4

Current Tools for Understanding Disease Distributions

Spatial epidemiology is a well-established field, as disease researchers and public health workers have been mapping disease occurrences for well over a century. The enterprise of mapping disease transmission risk, however, is far from simple or easy, and presents extremely complex challenges. The current toolbox for dealing with such questions is rich, in the sense that numerous methods and ideas have been developed, but it falls short because mapping efforts frequently fail to consider disease transmission in appropriate ecological and biogeographic frameworks. As a consequence, mapping results can be incorrect, imprecise, coarsely resolved, or incomplete, and geographic patterns of risk are not captured optimally.

Manifold factors mold the geography of disease transmission. Key causal influences include the population dynamics of multiple species; interactions among various species that may be positive, negative, or neutral; genetic, cultural, and social factors that modify patterns of infection and transmission; and the geographic complexities of the environment itself. The challenge of mapping transmission risk and anticipating occurrences of diseases across landscapes is not simple, and indeed can be quite demanding in analysis and inference.

Epidemiologists have been mapping disease occurrences for centuries. T. Koch (2005) presented maps of plague outbreaks in Europe that date back to the 1690s, suggesting that depicting the spatial dimensions of disease occurrence has long been a part of dealing with disease. In addition to plotting occurrences, the idea of interpolation began to be explored around 1840 in a map of sanitation in Leeds (T. Koch 2005). Using these two techniques, researchers were able to anticipate the potential for disease transmission in sites and areas where its occurrence was not yet documented, which was an important new possibility. Since that time, much energy has been invested in evolving and improving these interpolation methods to represent transmission risk optimally. This chapter attempts to summarize the current status of this toolkit that modern spatial epidemiologists use to map disease transmission risk.

The Current Toolkit
Occurrence Maps

The simplest approach to mapping disease transmission risk is still the technique explored more than 200 years ago by Seaman (1804), who placed dots on a map of yellow fever cases in New York City. In a more famous example, Snow (1855) used spatial information as a pointer by which to identify a contaminated well that was the source of a cholera outbreak in London. Snow suggested that the handle of the pump associated with that well be removed, rendering it useless, and the cholera epidemic tapered off quickly (T. Koch 2005). This approach—dots on maps—has the advantage of representing the original data clearly, allowing the reader/viewer to make her or his own interpretations (for a particularly elegant early example, see Maxcy 1928), and it still has considerable merit, at least in tandem with more powerful methods.

Techniques for creating such maps, of course, have improved markedly in the past 150 years. Two particularly important technologies have emerged. First, geographic information systems (GIS) permit ready exploration, visualization, and analysis in spatial dimensions. GIS platforms allow geographic summaries of occurrence data to be prepared in seconds, rather than the hours (or worse) that the process required just a few years ago. Second, satellite-based global positioning systems (GPS), which permit geographic referencing for any spot on Earth to 100 m or finer, allow precision in a step that formerly required guesswork, or at best involved considerably greater levels of spatial uncertainty. These two innovations have greatly facilitated the process of mapping disease occurrences.

Nonetheless, occurrence maps, a noticeably simple element in the current toolkit, have some serious failings. Although best guesses regarding occurrence points are plotted, the uncertainty around those points is seldom represented. The examples presented in Peterson et al. (2006a) and

Nakazawa et al. (2010) illustrate massive contrasts in the uncertainty associated with different occurrences of two diseases. One occurrence may be known and characterized with great precision, while another may have considerable uncertainty associated with it. The lesson to be learned is that disease occurrence data must be recorded in tandem with quantitative information on the degree of uncertainty in localization (Peterson 2008b), and this uncertainty should be expressed when disease occurrences are mapped.

Spatial Interpolations

When disease occurrence data are available as points, interpolations can be developed via kernel functions (for smoothing surfaces) or kriging (to extend predictions between points of known value). With these techniques, the density of points falling within windows of different sizes and shapes is used to smooth over broader areas than merely the points of known occurrence. This approach summarizes, in spatial dimensions, patterns of both disease occurrence and sampling, but it can also obfuscate details of variation in transmission risk as the radius of the averaging is increased.

Tools for spatial interpolation get more complex when disease occurrence data are reported at the level of polygons, often representing administrative units, such as counties (Pfeiffer et al. 2008). In conjunction with cases aggregated to administrative units, prevalences or standardized mortality/morbidity ratios become the target of mapping, instead of simply plotting occurrences and their density over space. These indices can be highly informative, as they can take into account population densities and other human population characteristics, thereby refining risk estimates and interpolations.

The most common method using spatially aggregated data is a Bayesian approach, which can incorporate both neighborhood relationships of occurrences and their associated spatial uncertainty (Pfeiffer et al. 2008). With newer hierarchical modeling approaches, Bayesian methods have impressive flexibility and power, including such features as different dependent variable structures, alternate error structures, and spatial autocorrelation (Lawson 2008). These approaches clearly bring impressive analytical power to questions of disease transmission risk mapping.

These analytical capacities, however, cannot get around certain cruel realities of the complexities of mapping disease transmission risk. For instance, spatial interpolations will always be sensitive to the spatial details of sampling, since the only information available to the model are the input data. Not only can artificial concentrations of sampling bias models, but details that do not get captured by the sampling available over space will be missed. Moreover, particularly in the case of spatially aggregated data, spatial resolution can be rather dramatically lost (e.g., Nakazawa et al. 2010).

The simple fact is that spatial interpolations have relatively little information available with which to fill gaps in sampling. The only data currently incorporated into such interpolations are (1) the occurrences or rates for the points or polygons sampled, and (2) the geometry of these points or polygons in space. More information can clearly be brought to bear on these challenges, which is really the heart of my arguments in this book.

Environment-Assisted Inferences

A rich source of additional information for enriching what is incorporated into disease transmission risk maps is the spatial pattern of environmental variation across the region of interest, which can include dimensions such as climate, land cover, substrate, and other relevant characteristics. Nonetheless, many books and texts on spatial epidemiology make no mention of these types of information. Of the few such texts that do refer to environmental covariates, statements in them are brief and variables are handled in a markedly nonecological context. Lawson (2008), for example, mentioned and explored "ecological" variables such as human socioeconomic status and poverty in a single chapter. Such approaches miss the point that many diseases are species that have physiological constraints on their distributional potential and ecological interactions with other species, and that all of these complexities occur across real-world geographic and environmental landscapes.

Recent publications in spatial epidemiology, disease ecology, and risk mapping have incorporated ecological variables more broadly. R. Eisen et al. (2007) used logistic regression models to seek associations between landscape features and human plague cases during 1957–2004 across a four-state region of the United States. Once models were evaluated (tested for predictive ability), the authors extrapolated their model rule sets to identify other areas at risk, which have perhaps not yet manifested actual cases, but hold appropriate conditions (that match those of sites where cases have occurred). These areas covered 14.4% of the four-state region. The authors presented elegant logic to back their assumptions that random control points represent an uninfected remainder of the human population; such assumptions are common in the ecological niche-modeling world (Peterson et al. 2011). Still, their analysis made assumptions in which sensitivity (omitting known occurrence points) and specificity (including other areas within the predicted area) were weighted equally, which allow known plague occurrences to be left out of the predicted area in order to avoid positing overly broad areas as suitable regions. These assumptions of error balance should be reconsidered in light of the known properties of positive and negative occurrence data (chapter 9).

Another recent example in the literature is a component-based (chapter 6) study of tularemia risk across Missouri by Brown et al. (2011). Because Missouri represents a focus of tick-borne diseases (e.g., spotted fever rickettsiosis, tularemia, and ehrlichiosis), the authors sampled ticks at 79 sites across the southern part of the state. They used logistic regression approaches, in tandem with a diverse set of potentially important environmental variables (land cover, Landsat imagery, 30–year climate averages, elevation), to establish that exposure to *Amblyomma americanum* nymphs or adults was greater in forested areas, and that high relative humidity in summer was another important factor. They also demonstrated significant positive associations between modeled suitability for the ticks and county-level case reporting rates for tularemia. As with the previous example, however, the methods that were used for model evaluation weighted omission and commission dimensions equally, and thus placed similar confidence levels in positive and negative occurrence data. This assumption is not realistic in a context of ecological niches and geographic distributions of species, particularly when spatial sampling is not comprehensive.

Overall, the incorporation of environmental information in models of disease risk assessment has been rather ad hoc in nature. The textbooks and treatises on spatial epidemiology have not included sufficient environmental information or substantive ecological thinking. Many individual researchers have seen the potential in such approaches, but they have used statistical tools to build inferences without the benefit of the thought frameworks founded in distributional ecology and biogeography, which can guide analyses toward appropriate conclusions regarding geographic distributions of pathogens and other disease-relevant species.

Shortcomings of the Current Methodologies

To emphasize the problems in the current suite of approaches, I explore their shortcomings in greater detail. The themes will be familiar ones, but they are sufficiently systemic—and intrinsic—in most work in this field that it is worth beating this dead horse a bit more. The problems on which I focus are consistently ones of lack of attention to data quality (chapter 8), the relative weighting of different error types (chapters 12 and 13), and nonbiogeographic thinking in processing and interpreting model outputs (chapter 13).

I will begin by exploring a relatively positive example (Glass et al. 2000), which focused on outbreaks of hantavirus pulmonary syndrome in the southwestern United States in the early 1990s. The authors used a case-control approach, where known positive sites were carefully matched with known negative sites to facilitate statistical comparisons. These occurrence (and nonoccurrence) data were analyzed in the context of a diverse set of environmental dimensions (rainfall, elevation, Landsat imagery). Although the authors used receiver operating characteristic (ROC) approaches to model evaluation, which weight positive and negative occurrence data equally (chapter 14), they explored the ROC plots in ways that emphasized correct prediction of presences over correct prediction of absences. Thus the authors avoided many of the pitfalls of assuming that all errors are created equally.

A contrasting example is that of Thomson et al. (1999), who sought to outline the geographic distribution of *Phlebotomus orientalis*, the sandfly vector of visceral leishmaniasis in Sudan. Their study was based on presences and absences of the species at 44 collecting sites across central Sudan. A logistic regression model was used to estimate the probability of presence at each collecting site in relation to climatic and environmental variables. The occurrence data, however, showed a strong spatial separation of positive and negative occurrence sites (figure 12), which leads to situations in which the species' responses to environmental variation are not observed under conditions that are extreme or nonoptimal for the species (Owens et al. 2013). As a consequence, it is highly doubtful that full response profiles were established for the species with respect to the independent variables. Furthermore, the model outputs were masked by the "known" rainfall-based boundaries of the distribution of *Acacia-Balanites* woodlands, which most likely hides a lot of detail about the true distribution of the species. Worst of all, no attempt was made to evaluate the models, which could have been possible by assembling independent occurrence data sets from other sources (e.g., collections in natural history museums) to document the distribution of this important vector species.

Still more problematic are the analyses presented by Fichet-Calvet and Rogers (2009), which aimed to create a risk map for Lassa fever infections in humans across West and Central Africa. The authors assembled a rich set of descriptors of environmental conditions throughout the region (from Scharlemann et al. 2008), which should have provided a basis for detailed model predictions, and they used nonlinear maximum likelihood discriminant analysis to create their models.

The only model evaluations they provided, however, were based on the kappa statistic, which is a performance measure and not a meaningful test. Moreover, kappa not only weights presence and absence data equally, but also depends on the absence data, which for this study were resampled randomly from areas close to known occurrences. The authors resampled

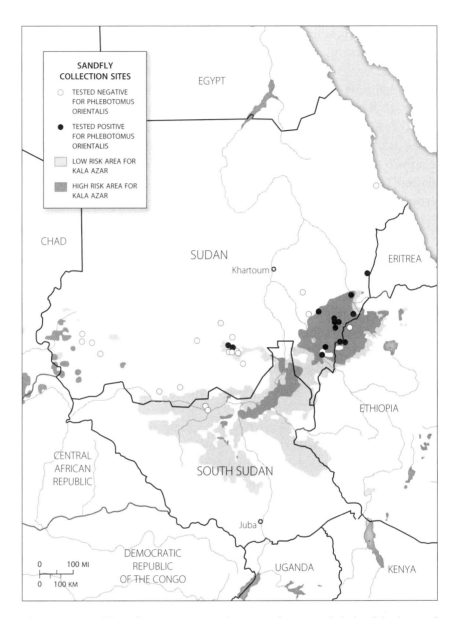

Figure 12. Sandfly collection sites in the central savannah belt of Sudan and South Sudan. Adapted from Thomson et al. 1999, © 1999 John Wiley and Sons.

1,000 points in this way, but they could have used 100 or 1,000,000 such points—the effects of the number of resamples on the kappa statistics would have been dramatically different, but the authors used the statistics nonetheless as a simple way of concluding that their models were excellent.

More fundamentally, these authors assembled occurrence data for the virus but did not (apparently) inspect their data with regard to the level of confidence in the diagnosis. An extreme geographic outlier placed the virus in the Republic of the Congo, well south (~1300 km) of the known range of the disease, based on a single study (Talani et al. 1999) that could easily have been compromised by false-positive errors. A simple eyeballing of the data set spatially and a careful reassessment of the evidence backing up the spatial outliers would have detected and potentially removed this suspect occurrence data point, as well as several others. Moreover, the authors paid no attention to the massive oversampling in Sierra Leone and adjacent sectors of Liberia and Guinea that resulted from intensive, active surveillance efforts there, in stark contrast to areas elsewhere in West Africa. Finally, the analytical methods they used appear to weight omission and commission errors equally, so that including one of the 1,000 random absence points within the predicted area is as serious an error as leaving out a known transmission site, even though active surveillance has never occurred on the vast majority of the landscape. The symptoms of all of these problems are apparent in the maps they presented: (1) likely overconcentrations of points in Sierra Leone and Nigeria, (2) outliers to the north and south that are not included within the area predicted as being at risk by the model, (3) modeled suitable areas that mainly correspond only to the areas sampled, and (4) disjunct suitable areas separated by areas that do not seem particularly distinct environmentally. These concerns and alternative risk maps are presented in Peterson et al. 2014.

Conclusions

The current toolkit available to spatial epidemiologists includes several interesting and potentially powerful tools. In particular, the methods that have evolved for managing complex spatial occurrences hold considerable possibilities for analyzing spatial manifestations of disease phenomena (Waller and Gotway 2004). Efforts by spatial epidemiologists to include environmental covariates in models, however, have differed in their quality and rigor.

Were these questions to be recast and reconsidered in the context of ecophysiology, distributional ecology, and biogeography, the same tools most likely could be put to sound use. Lessons from ecophysiology and distributional ecology indicate that spatial position will be less important than the conditions manifested at the sampled sites (Peterson 2003), and biogeographic thinking will be of further assistance in bringing real-world geographic features (e.g., barriers to dispersal) into the picture. With these biological frameworks in mind, the current toolkit can be used to maximum benefit, and potentially new tools can be incorporated in situations where they offer additional insights.

PART II

DISEASE MODELING BASICS

5

Modifications to the Basic Framework

Disease transmission systems show considerable complexity that is not manifested as dramatically in other elements of biodiversity. In particular, interactions among species are frequently of extreme importance, and they can drive much of the spatiotemporal dynamics of disease transmission. These interactions can be highly complex, with multiple lineages of pathogens being maintained and transmitted by communities of hosts and vectors. These species-to-species interactions are then potentially modified by factors such as immunity or health status, with important implications for disease transmission geography.

Peterson et al. (2011) offered the Eltonian Noise Hypothesis as a means of reconciling the possibly important but nearly intractable effects of interspecific interactions on ecological niche models and the need for geographic views of species' distributions. The Eltonian Noise Hypothesis suggests that biotic interactions are frequently diffuse and nonspecific, and that they are also generally manifested at spatial resolutions so fine as to make them nearly irrelevant to the macrogeographic viewpoints that are the focus of niche modeling. Considerable evidence can be marshaled to back this hypothesis, such as the ability of niche models trained on one continent to anticipate the distributional potential of an invasive species on another continent where the suite of interactor species is often quite different (Peterson 2003).

Diseases, however, frequently present interactions that can have macrogeographic implications, such that the Eltonian Noise Hypothesis most likely would fail. In some cases, interspecific relationships are long-term, coevolved relationships that appear to be very tight (Nemirov et al. 2004). In others, one or a small group of vector species is responsible for the entire geographic range of the pathogen, and, without the vector, the pathogen would not be transmitted (Peterson 2007a). Hence, in disease biogeography at least, the Eltonian Noise Hypothesis cannot be taken as a given, and special attention must be paid to a broad diversity of interspecific effects, which are the subject of this chapter.

Disease Peculiarities

A large body of models and theory in epidemiology has developed into a detailed understanding of factors that should affect disease transmission. The Susceptible-Infected-Recovered (SIR) framework offers a simple, nonspatial starting point for considering these additional important factors (Keeling and Rohani 2008). Susceptible individuals are infected by other individuals who already have the disease and both will either die or recover. Transmission can cease, owing to too few individuals being infective, or too many individuals having recovered (and who therefore become immune), which can produce either stable equilibria or density-dependent oscillatory behavior. Such cyclic outbreaks are observed in some viruses that are periodically reintroduced into regions by migratory birds, and SIR dynamics offer a useful platform for contemplating these cycles (Day 2001; Zeller and Schuffenecker 2004).

A diversity of host-related and vector-related factors also enters the picture (Keeling and Rohani 2008). Demographic structure, latent periods of infection before infectivity, carrier states, behavioral diversity that creates different risk groups, preexisting immunity, inherited immunity, genetic composition, and nutritional status all may change disease dynamics significantly. Interactions among pathogens are also known, where infection with and recovery from one pathogen may confer full or partial immunity to infection by another pathogen (Tesh et al. 2002). On the other hand, sometimes infection with one pathogen may increase susceptibility to infection by another (Kurane and Ennis 1992; Keeling and Rohani 2008).

In sum, it is clear that simple and plain niche-modeling approaches will not be sufficient to estimate the transmission risk of many diseases. Although one could conceive of these manifold factors as the details that shape **B**, they add considerable complexity to the overall challenge of estimating the geography of disease transmission (Peterson 2008a). Niche models can provide a starting point: a summary of the distribution-environment relationships for individual components of transmission systems (chapter 6). From these building blocks, better and clearer views of transmission risk can be built by including what is known about these diverse modifying factors, generally via scenario building or simulation in a GIS environment (chapter 15).

Real-World Examples: West Nile Virus and Others

West Nile virus was discovered in the West Nile District of Uganda in 1937; although additional cases were detected in subsequent years, it was not recognized as the cause of severe human disease until a 1957 outbreak in Israel. This coupling was augmented still more with major but scattered and unpredictable outbreaks across southern Europe and the Middle East in recent decades. No appreciable avian mortality was noted until a 1998 outbreak in Israel, where significant avian die-offs were documented (Lanciotti et al. 1999), possibly as a result of evolved virulence. Finally, West Nile virus found fame when, in 1999, it jumped over the Atlantic Ocean by unknown means and became established across the Americas, spreading initially from the New York City area, then reaching the Caribbean before 2003 (Dupuis et al. 2003, 2005; O. Komar et al. 2003), and Argentina by 2006 (Morales et al. 2006).

Curiously, although avian mortality was widespread across the United States, and many human and equine cases were documented, such evidence was much less notable as the virus progressed southward. Considerable speculation has centered on these differences, although it is possible that they are merely apparent and result from differences in diagnostic abilities and the reporting of avian mortality. A sizeable diversity of closely related viruses is present across the region, however. A 2002 survey of serological reactivity in the Dominican Republic found five seropositive samples for West Nile virus, but four samples were reactive to a more generic flavivirus antigen and not to West Nile virus (O. Komar et al. 2003). The implications of this background of potential interspecific interactions cannot be neglected in efforts to map disease transmission risk.

Detailed experimental work has revealed considerable complexity. Brault et al. (2004) documented complex cross-immunity reactions and host-competence shifts among different strains of West Nile virus. Several researchers have suggested that the presence of other flaviviruses may dampen the potential for major outbreaks in Latin America (Estrada-Franco et al. 2003). Latin America has dengue, St. Louis encephalitis, Ilheus, Bussuquara, Jutiapa, and yellow fever viruses, unlike the United States and Canada, where exposure to other flaviviruses is much more limited. Previous exposure to other flaviviruses could attenuate West Nile infections, owing to a cross-protection from antibodies, or could possibly worsen them by enhancing susceptibility to the contagion. Nonetheless, careful experimental and epidemiological studies suggest that the spread of West Nile virus is unlikely to be halted significantly by previous immunity to St. Louis encephalitis virus (Brault et al. 2004; Mackenzie et al. 2004).

West Nile virus is far from the only possible example. Lyme disease presents a fascinating example of nonlinearities and challenges to a simple BAM framework. The disease occurs across the northeastern United States, and then in a disjunct area in the northcentral region. Both the disease and its principal vector, the tick *Ixodes scapularis*, show a curious area of absence, centered on the state of Ohio (Brownstein et al. 2003), that is without known climatic correlates (Ashley and Meentemeyer 2004). Some environmental parameter that is not being considered in the models could account for this disjunction; or the entire system could be out of equilibrium, such that the holes will fill in over time (Hoen et al. 2009); or some intriguing interspecific interaction could be involved (Ostfeld et al. 2006).

Yellow fever offers another intriguing case. This disease apparently originated in Africa but was transported to the Americas before 1700, most likely with slaves or mariners (Barrett and Higgs 2007). The virus is transmitted in sylvatic primate-mosquito cycles on both continents, and in the Americas it is transferred to humans chiefly by the mosquito *Aedes aegypti* (Barrett and Higgs 2007). Curiously, this virus has never emerged in Asia (Monath 2001), although potential vectors appear to be present there, and Southeast Asia is thought to be vulnerable to the emergence of yellow fever. Assuming that colonization of Asia has been possible, with infected persons having been bitten by appropriate mosquito species, explanations for the absence of yellow fever there include cross-protection from antibodies to dengue, and a low vectorial competence of local mosquito populations (Monath 2001). Another possibility might include the absence of key species of vectors or hosts through which to establish a sylvatic cycle, although transmission did occur in North America, where many of the primate species that host the virus elsewhere are lacking. Regardless, the answer to these questions does not lie in simple associations with climate or other abiotic parameters (Peterson 2008a).

Implications for Disease Modeling

The elements treated in this chapter are neither simple nor straightforward to incorporate in the process of mapping disease transmission risk. Nonetheless, their influences are clear, and they must be considered directly, or large-scale inaccuracies can result. Ashley and Meentemeyer (2004) modeled climatic associations of Lyme disease cases across the United States, but they included *only* climatic factors; the risk maps that resulted were rather general and did not reconstruct the range disjunction discussed above at all. Other factors influencing disease transmission geography are much more subtle, making the steps discussed below important.

A first crucial point is model calibration, to avoid biased estimates of ecological niches of key species. For instance, most hantavirus transmission is focused in the Four Corners region of the southwestern United States, yet the virus has a broader known range; apparently, the Four Corners focus results in large part from the socioeconomic status of human populations in the area, which permits closer proximity to the rodents that host the virus. As will be discussed in chapter 6, these filters can affect one's view of the transmission geography and ecology of the phenomenon, to the point where models are biased or erroneous if all contributing factors are not pondered carefully.

The major instance where these factors are important, however, is in estimating the risk of transmission. When influences are rather simple and are based on distributions of other species, such as in cross-immunity among viruses, the distributions of these other species can be directly incorporated into analyses by linking several ecological niche models. Although each model will present multiple challenges to assembling sufficient occurrence data and controlling confounding factors, the additional information from the distributional patterns of these other species can be quite informative in mapping disease transmission risk. Host and human/animal characteristics (e.g., socioeconomic status, immunity, etc.) are best incorporated at the stage when the risk map is assembled, which occurs after the basic ecological characteristics are estimated. (These points will be treated in chapter 15 in much greater detail.)

Finally, for interspecific effects, once the basic geographic-environmental ties are estimated and characterized—in other words, when the niche models are calibrated—it is important to bear in mind that these models only outline distributional potential. Unless interactions are quite strong, direct, and manifested consistently in environmental dimensions, which would bias the niche estimates, niche models should offer a picture of possible distributional areas without including the effects of complex interactions. These interactions can then be incorporated in the risk assessment stage (chapter 15).

Conclusions

This chapter has been one of caveats. Niche models summarize distributional potential, but that is not the whole picture. Details of variations in host populations and of the dynamics of interspecific interactions can change the geographic picture of disease transmission considerably. As a result, researchers must be cognizant of these complexities, avoid allowing biases to enter at the niche estimation stage, and incorporate these factors carefully in estimates of transmission risk. The end result is a picture of transmission risk that is as complete and unbiased as possible.

6

Modeling Components versus Outcomes

The realities of the world of diseases and the data that document their geography present two distinct types of information and analyses that can—in theory at least—be used to estimate geographic patterns of transmission risk. When biological actors (e.g., pathogens, vectors, hosts, etc.) are involved in a transmission system, one can either build a view of risk from the components of the transmission system or use the end result of the transmission system (e.g., cases of infection in humans); the latter can be called a "black-box" approach. Each of these options has advantages and disadvantages, but frequently only one will be possible in a given situation, because data availability imposes limitations. When possible, however, evaluating and comparing the two methods offers exciting potential insights into the spatial and ecological filters that make them similar or different.

Waller et al. (2007) provided a useful conceptual framework for contemplating disease transmission systems and their observable features (figure 13). The leftmost part of this figure shows the different dimensions that characterize the environment in which the species is distributed; next, the figure shows four maps that correspond to distributions of vector and host species. The combination of these four components of a hypothetical transmission system builds a view of the true incidence pattern (the middle map), which in this book is referred to as a "component-based"

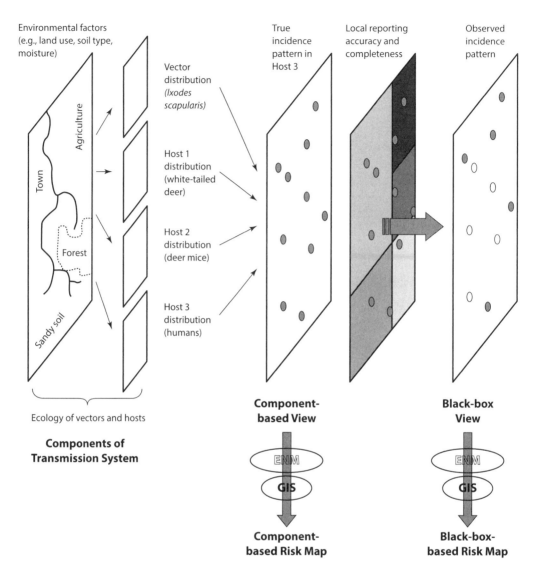

Figure 13. Conceptual elements in the development of risk maps for disease transmission, adapted from a diagram regarding Lyme disease transmission. See the text for a detailed explanation. From Waller et al. 2007, with kind permission from Springer Science and Business Media.

picture of disease incidence patterns, derived from individual estimates for each actor in the system.

This true incidence pattern is constructed from the components of the transmission system. If one or more of the actors were either unknown or, for practical reasons, not included in the picture, the estimated pattern would change in ways that would not be clear or easily anticipated. Hence the component-based picture approximates the true incidence pattern only to the extent that all of the pieces are in place and are correctly estimated.

Waller et al. (2007) then included in their schema a map of "local reporting accuracy and completeness," which is effectively a filter that modifies the true incidence pattern into what they termed the "observed incidence pattern." This filter is a composite of all of the factors that may modify incidence patterns: the level of knowledge about the disease by medical staff, the availability of diagnostic technologies, the accuracy of the diagnoses, transcription errors, the kinds of reports submitted to central data centers, and so forth. (See figure 11, in chapter 3, for another way to visualize these ideas.) The observed incidence pattern is thus the result of the filter modifying the true pattern. Occurrence data that are available to characterize this observable pattern are what is referred to in this book as "black-box" data, because the researcher generally does not have any idea of what the filter layer looks like.

This chapter explores both component-based and black-box approaches to mapping disease transmission risk. Related ideas have sometimes been proffered in contrasts between knowledge-driven transmission risk models (a first-principles approach based on components) and data-driven models (more closely akin to the black-box models discussed in this chapter). Stevens and Pfeiffer (2011) presented a discussion of these ideas. As the reader will see, each type of has strengths and weaknesses that, in particular situations, may make one method preferable over the other. The most desirable approach is to use both and compare them, thereby extracting an estimate of the reporting filter, which turns out to have considerable interest in and of itself (Waller et al. 2007).

Disease Transmission Systems as Sets of Interacting Species
Examples of Transmission Systems

Disease transmission systems are complex, representing a series of interactions among a diversity of species. At the simplest end of the spectrum are environmental diseases, such as the soil fungi *Blastomyces dermatitidis* and *Coccidioides* spp. These pathogens are maintained in the soil, and they can be transmitted to humans when the soil is disturbed and people breathe in the resulting dust. Such pathogens have very simple transmission systems, in that the only questions are of whether soil conditions are appropriate for maintaining the pathogen, and whether environmental conditions are adequate for infecting humans or other animals. Note, however, that no biological actors appear to be involved in the transmission system, other than the pathogen itself.

The next level of complexity is that of pathogens that have an animal

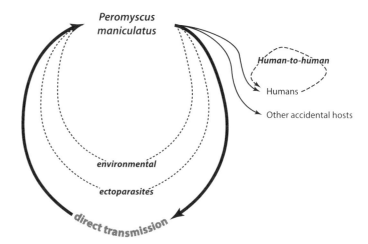

Figure 14. A summary of likely and possible transmission pathways for Sin Nombre virus.

host, but are transmitted directly among hosts or from the host to humans (or human-associated animals). A particularly well-studied example is the hantavirus called Sin Nombre virus, detailed in figure 14. The virus is held in a reservoir that is linked to the field mouse *Peromyscus maniculatus*. The principal means of maintaining this virus in mice is via direct transmission of the virus from mouse to mouse (Mills et al. 1999). Other possible routes by which viral infections could be transmitted among and maintained in rodent populations include environmental transmission (Kallio et al. 2006; note, however, that this paper refers to a European hantavirus only) and transmission via ectoparasites (Houck et al. 2001). The possibility of human-to-human transmission (Enria et al. 1996; Wells et al. 1997a) does not appear to be likely for Sin Nombre virus (Vitek et al. 1996; Wells et al. 1997b), although it is suspected for other hantaviruses (Wells et al. 1997a; Padula et al. 1998). In this case, transmission via vectors appears to be unlikely or negligible; the only organism involved that has an outward, exposed life history phase (other than the virus itself) is the mouse, providing a clear target for modeling efforts. Other disease systems of this sort include Ebola, Marburg, and Lassa fever viruses.

The complexity of transmission systems then grows rapidly. The subsequent level adds a vector to the host and the pathogen. One example is West Nile virus, where avian hosts are linked via mosquito vectors (figure 15), so the system consists of three general sets of interacting species: pathogen, host, and vector (Marfin and Gubler 2001; Petersen and Roehrig 2001). Moreover, both the hosts and the vectors appear to include multiple species. Among birds—where the most competent reservoirs appear to

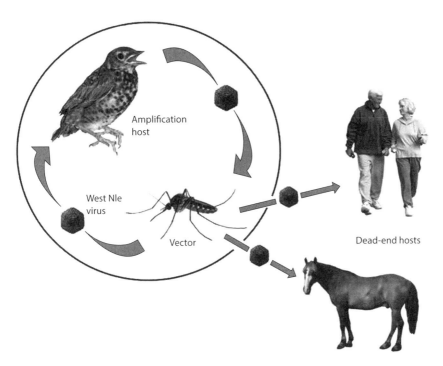

Figure 15. A summary of potential transmission cycles for West Nile virus, showing hosts and vectors, as well as two example dead-end hosts. Adapted from www.vetmed.wisc.edu/WNV/background.html.

be nonmigratory songbirds (N. Komar et al. 2003)—the pattern of spread cannot be adequately explained without also considering the movements of migratory bird species (Peterson et al. 2003). In addition, among mosquitoes, it appears that both the bird-biting ones (*Culex* spp.) and what have been termed "bridging vectors" (e.g., *Aedes* spp. and *Ochlerotatus* spp.) that have more catholic tastes may be involved in the overall system (Kilpatrick et al. 2005; Turell et al. 2005; Hamer et al. 2008), and that different mosquito species were implicated as the virus's range expanded (Sardelis et al. 2011). Hence the complexity of this system is massive in comparison with the previous group of diseases.

Perhaps the best illustration of maximum complexity in a disease transmission system structure is that of plague (figure 16). Here the totality consists of multiple, interlinked transmission systems: an epizootic and highly unstable cycle in *Cynomys* prairie dogs (Cully et al. 1997), a poorly characterized enzootic cycle in some other small mammal species or set of species, and transmission to and from (and perhaps among) carnivores (Salkeld and Stapp 2006), not to mention occasional human-to-human

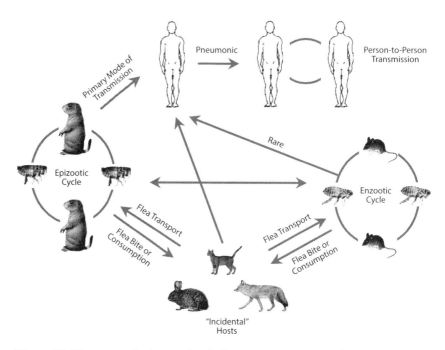

Figure 16. The transmission cycle of plague in western North America. From Gage and Kosoy 2005.

transmission in the form of pneumonic plague. Actors in the plague cycle thus include small mammals, medium-sized carnivores, fleas, and humans. The overall complexity is enormous, creating a system that is especially difficult to evaluate and capture in models (Girard et al. 2004; Webb et al. 2006; Maher et al. 2010).

Examples of Contrasting Ecologies of Interacting Species

As mentioned in chapter 2, ecological niches represent inherited and evolved features of lineages. Peterson et al. (1999) were the first to survey patterns of change in coarse-resolution ecological niches across a broad suite of species, concluding that niche conservatism characterized most evolving lineages. This early result was corroborated by a careful reanalysis (Warren et al. 2008). A detailed meta-analysis that covered 76 publications and 299 lineages also showed widespread niche conservatism, at least over time scales of 10^5–10^6 years (Peterson 2011). In general, ecological niches of evolving lineages may be expected to diverge slowly; on the other hand, if all life is related (a monophyletic lineage), niche divergence has clearly occurred broadly across the tree of life on Earth.

To illustrate the divergence of ecological niches over time, figure 17 shows the separation of ecological niches across one sublineage of *Pero-*

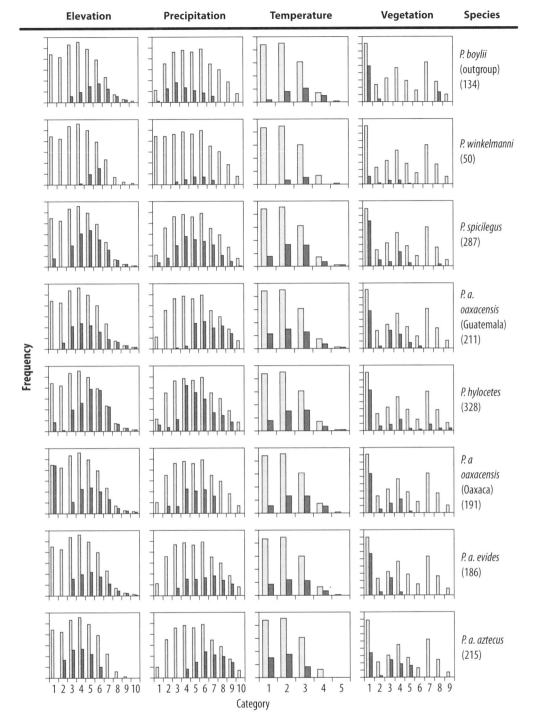

Figure 17. Modeled distributions of eight *Peromyscus* taxa with respect to four environmental dimensions. Empty bars represent the availability of the individual environmental categories in Mexico; filled bars represent environments in which each taxon is predicted to occur. Numbers below the taxon names indicate the quantity of unique environments (niche breadth) that each taxon was predicted to inhabit by three or more of the five models. From Martínez-Meyer 2002.

myscus mice (Martínez-Meyer 2002), which are probably the hosts of a parallel lineage of hantaviruses. That study assembled large data sets characterizing known occurrences of each species in the lineage and used niche-modeling approaches to relate those occurrences to landscape and climate features. Clear differences among species within this lineage can be seen, particularly with respect to temperature and precipitation levels, even though broader sets of conditions across the study region are available to the species. In this case, given that certain species have access to but do not colonize areas presenting the environmental conditions used by other species, conclusions of niche differentiation are likely to be robust (Barve et al. 2011).

A recent study of plague (Maher et al. 2010) illustrated the complexity of niche dimensions among elements in a disease transmission system even more plainly. Plague has a clearly delimited range in western North America, with the eastern edge of its post-1920 distribution termed the "Plague Line" (Antolin et al. 2002), and a dramatic focus located in the Rocky Mountains and the Four Corners region. Maher et al. (2010) tested two hypotheses for why plague is distributed where it is and why it is not present elsewhere: (1) a specific niche for plague exists (the "plague niche hypothesis"), perhaps determined by flea ecology and distributions, which were not included in the analysis, versus (2) plague ecology and distribution are consequences of the suite of ecologies and distributions of its hosts (the "host niche hypothesis"). They tested the coincidence of niches between the overall distributions of eight plague host lineages and plague distributions and found it to be almost nonexistent (figure 18, bottom half). For those same eight hosts, however, the sites where the hosts were documented as being infected with plague were highly coincident environmentally with plague distributions (figure 18, upper half). The conclusion was that a plague niche is indeed what determines the geographic distribution of plague in North America, which contrasts with the niches of many of the hosts of plague in the region. Put another way, mammal hosts are simply hosts of convenience for plague, and plague distributional ecology appears to be driven more by vector ecology. In the context of the present chapter, this study serves to illustrate the multifactorial and highly complex nature of disease transmission cycles in nature.

Black-Box Approaches

The simplest way to get started with modeling ecological niches and estimating geographic distributions of a disease system is to analyze only the geographic distribution of cases of the disease. This approach will be referred to as a black-box approach, because it effectively wraps up all of the

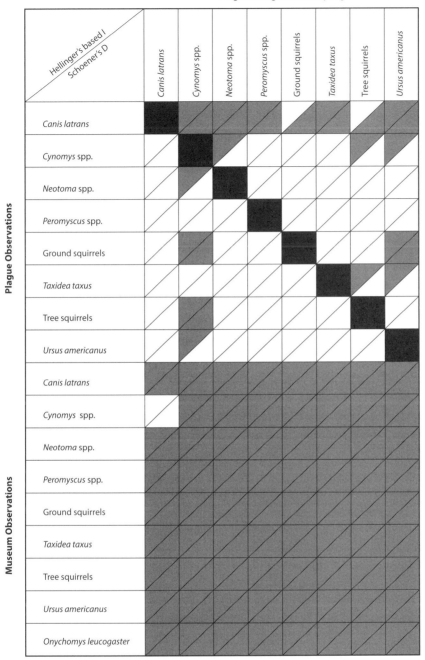

Figure 18. Results of niche similarity testing for hosts of plague in western North America. *Top panel*: records of species infected with plague. *Bottom panel*: any record of the host species. White cells indicate tests in which similarity could not be rejected; gray cells indicate tests in which similarity was rejected. From Maher et al. 2010.

complexity of the individual ecologies of each of the components of the disease transmission system into a single response variable. A black-box approach has both advantages and disadvantages: a single response variable is much easier to analyze than the independent ecologies of numerous participating species in a particular transmission system, but sometimes no option is available. For many diseases, knowledge of the transmission system is either nil or only anecdotal or partial. In such cases, the approaches detailed in this book can be used to identify, or at least characterize, potential participants in the transmission system (chapter 18).

Black-box approaches and single, overall response variables nonetheless incorporate other factors into the modeling picture that may or may not consistently pertain to transmission risk. If the models being built are made up of distribution-environment connections (niches), the distribution of disease cases in environmental dimensions may differ distinctly from where the pathogen actually occurs. In the case of hantaviruses, Sin Nombre virus is tightly associated with the mouse *Peromyscus maniculatus*, which is distributed across the entirety of the United States, yet the great bulk of known infections of this virus occur in the western United States, particularly in the southwest (Monroe et al. 1999). Nonetheless, the virus does occur across the continent. On the other hand, in residential districts of eastern port cities in the United States, nearly 50% of Norway rats (*Rattus norvegicus*) are infected with hantaviruses, yet confirmed cases of infection are negligible or nil (Yanagihara 1990). Perhaps most strikingly, no human cases of Hantavirus Pulmonary Syndrome have been documented in Mexico (Sánchez-Cordero et al. 2005), in spite of emerging evidence that the virus (Suzán et al. 2001; Chu et al. 2008; Castro-Arellano et al. 2009; Milazzo et al. 2012) and its hosts (Hall 1981) are both widespread and diverse in that country. The picture that emerges indicates many more cases north of the border than south of the border, yet the implication is that some filter—probably related to diagnosis and reporting—dramatically reduces the number of incidences registered in the south. Similar reporting imbalances for hantaviruses apply to African hantaviruses (Bi et al. 2008).

Examples of analyses of disease systems based on black-box case distributions are probably best developed under three circumstances. First, in many cases knowledge of the transmission system does not permit any alternative. For example, until recently the host of filoviruses remained completely unknown, and the sole source of knowledge about their distribution was via the location of human cases. As a consequence, the only option available for risk mapping was that of plotting known cases in humans (Peterson et al. 2004a, 2006a). Additional diseases with a poorly known natural history that precludes any analysis except black-box approaches include Buruli ulcer (Williamson et al. 2012), nodding disease (Winkler et al. 2008), and many others, particularly in tropical regions.

A second circumstance under which black-box approaches are indicated is when the transmission system is so complex that component-based models would be either impossible to create or overly simplistic if one wishes to capture the full behavior of the system. As mentioned above, West Nile virus is hosted in North America by a wide spectrum of bird species and is vectored by a diversity of mosquito species (Kramer et al. 2008), to the point where a component-based approach to this pathogen would be essentially impossible. As a consequence, efforts to map risk for West Nile virus have been limited to basing them on human case distributions or have required tightly constraining assumptions. Tularemia (Nakazawa et al. 2007, 2010) and avian influenza (R. Williams et al. 2008; R. Williams and Peterson 2009) are among other disease systems that are similarly complex and necessitate analyses at the level of human cases, rather than the full complexity of actors in their transmission systems.

The third situation under which black-box approaches are indicated is perhaps the most interesting and exciting. Here, these approaches are implemented and compared with component-based approaches to gain much greater insight into the overall transmission system. This set of ideas is treated later in this chapter, in the Combined Approaches section.

Component-Based Approaches

In relatively simple transmission systems, several analyses have developed vector-based or host-based predictions of risk. For dengue, which depends only on *Aedes* mosquito activity (in terms of species other than humans), a detailed analysis of monthly distributions across Mexico was able not only to predict spatiotemporal distributions of the vector species with significant accuracy, but also to anticipate the spatiotemporal distribution of dengue cases through the course of a year (Peterson et al. 2005). Other mosquito-based analyses include those focused on malaria (Levine et al. 2004a, 2004b; Benedict et al. 2007). Host-based analyses in such simple systems have included projections of likely risk areas for hantaviruses in Mexico (Sánchez-Cordero et al. 2005), where these pathogens remain largely unknown (Bi et al. 2008).

Modeling all of the pieces of a more complex disease transmission system and then assembling these components into a subsequent cohesive predictive model has not been possible in many situations, owing to the compounded challenge of putting together many data sets and models. In the plague example (Maher et al. 2010), while detailed human case-occurrence data were available, thanks to the hard work of a colleague (K. Gage, Centers for Disease Control and Prevention), the host occurrence data that were stored as paper files in 12 filing cabinets was able to be captured digitally (taking two years of work); only after this effort could any modeling occur. The extant data on distributions of flea species in North America have apparently not been digitized, or at least they have not been made easily available, so a full, component-based analysis of plague in North America must await further progress in basic data capture and management.

Three examples of component-based analyses have been developed, albeit each with a purpose oriented far more toward understanding transmission systems than toward actually mapping transmission risk. An early study of the spread of West Nile virus in North America (Peterson et al. 2003) built scenarios of how the disease might spread, based on distributions of key mosquito species and nonmigratory bird species, versus scenarios that included breeding and wintering distributions of migratory bird species (figure 19). The basic result was that only the latter scenario could explain the observed pattern of the virus's spread, thus providing strong evidence that migratory birds played a role in the hemispheric expansion of this virus, which occurred over the decade following its arrival in the Americas.

Another early study, examining host relationships of the *protracta* group of *Triatoma* bugs that transmit Chagas disease in Mexico, tested for host specificity in distributions of the bugs and the *Neotoma* woodrats that apparently host them (Peterson et al. 2002b). The researchers assembled occurrence data for each species of bug and woodrat, developed ecological niche model–based maps to summarize likely distributions of each species, and then evaluated patterns of range overlap between host and parasite species. The study found what appears to be a one-to-one correspondence between bugs and woodrats, to the point that a previously unknown host-parasite relationship could be posited: *T. barberi* was predicted to be a parasite chiefly (or perhaps exclusively) of *N. mexicana*. This picture of a tight host association in one group of Chagas vectors stands in contrast to apparently looser associations in other triatomine groups. A third example, of a partial, component-based study of plague transmission in western North America, has already been described above (Maher et al. 2010).

Combined Approaches

When detailed data on both the components of a disease transmission system and the end results of its cycling (cases of the disease in humans or other animals) can be assembled, the potential for deeper insights

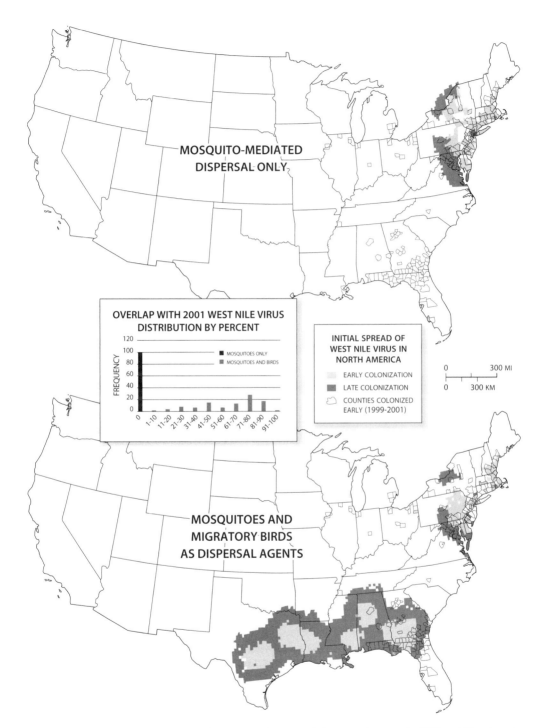

Figure 19. A summary of the results of simulations of the initial spread of West Nile virus in North America, based on assumptions of mosquito-mediated dispersal only, and mosquitoes and migratory birds as dispersal agents. From Peterson et al. 2003.

emerges. Unfortunately, such data-rich systems are rare. Most frequently, data for at least some of the components of the system will be problematic to assemble.

Consider figures 11 (in chapter 3) and 13 (in this chapter). The BAM diagram in figure 11 is cast in geographic dimensions, with the various intersections in the diagram referring to different distributional areas (Soberón and Peterson 2005; Peterson et al. 2011). Barve et al. (2011) presented a discussion on how to map **M**, and mapping **A** is the main object of niche-modeling exercises (Peterson et al. 2011). Mapping **B** is less straightforward, and some have even argued that it is largely irrelevant to geographic questions (Soberón 2007). Even those studies that have measured and mapped individual dimensions of **B** (e.g., Heikkinen et al. 2007) have not captured the full picture of biotic interactions. The point is that the entire BAM scenario (the bottom three bars in figure 11) can, at least under certain assumptions, be characterized in both geographic and environmental terms. BAM diagrams and models for different component species (when multiple components exist) can be combined to produce a view of the transmission patterns of the disease, but this approach will be effective only if transmission patterns depend solely or chiefly on the components that are included in the analysis.

The black-box approach, on the other hand, offers a view that integrates all of the factors shown in figure 11 (and others, if they exist). In other words, all of the various niches (environmental biases associated with a distribution) fold into a single, overall view. This concept is equivalent to estimating an ecological niche that summarizes environmental biases integrated over the whole phenomenon.

With these two approaches in hand, the contrasts between them become of great interest (Waller et al. 2007). If the component-based view of the disease transmission system is robust, then the difference between the black-box view and the component-based one basically offers a spatial characterization of the combination of the upper horizontal bars in figure 11. A spatial presentation of difficult-to-quantify elements (e.g., a reporting bias or a diagnosis bias) is rare, but it can be immediately informative. Biologically, comparisons between the two approaches may also point to missing factors in the component-based view, which can educate the model's structure concerning the basic transmission system.

As a first (partial) example of this process, González et al. (2011) compared the union of the distributional areas of the vector components of leishmaniasis transmission in Mexico with the overall distribution of cases of the disease. The sandfly *Lutzomyia olmeca olmeca* is the only proven host-to-humans vector of the parasite *Leishmania mexicana* in that country, so the researchers assembled occurrence data for this species and developed detailed models anticipating its geographic distribution across Mexico. The likely distribution of *Lu. o. olmeca* does not cover all of the reported cases of cutaneous leishmaniasis in the country, however, suggesting that additional species (e.g., *Lu. cruciata* and *Lu. shannoni*) are probably also involved in the transmission cycle. Similar reasoning was used to eliminate potential host mammal taxa from consideration in studies on the ecology, distribution, and host associations of filoviruses (Peterson et al. 2004a, 2004b, 2006a, 2007b; Lash et al. 2008). Avian influenza would be another example, except for the fact that component-based studies could be developed only in North America, because of data availability issues (Peterson et al. 2007a; Peterson and Williams 2008), while the black-box approach could only be implemented in Europe, Asia, and Africa, where strain-specific data exist to document large numbers of human or animal cases (R. Williams et al. 2008; R. Williams and Peterson 2009; Bodbyl-Roels et al. 2011; R. Williams et al. 2011).

Conclusions

Disease transmission systems and the partial data available to characterize them in terms of geography and the environment present significant challenges in assembling appropriate and sufficient data for risk mapping and niche modeling. Some systems lend themselves well to building up a picture of likely transmission patterns from their components, while others would be much better as an overall integration of the various parts across the whole system. When both types of approaches can be utilized, the result can be a rich set of insights into aspects of the transmission system that would otherwise be difficult to perceive and characterize.

7

Space-Only versus Space-and-Environment Models

This chapter focuses on the differences between the approaches outlined in this book and much of the present activity in the field of spatial epidemiology. Current analyses in this discipline frequently are developed in spatial dimensions only and, as a consequence, lose much of the information that could be revealed in environmental covariates. The sampling of disease-related phenomena is rarely (if ever) comprehensive, so even in the best of situations, some degree of interpolation becomes necessary. The question is how to achieve that interpolation: seeking patterns in space only, or incorporating environmental variation in the picture as well? If sampling is not extensive and intensive, or if sampling is biased spatially, the additional information provided by environmental covariates offers considerable advantages.

The science of epidemiology has developed a rich set of heuristics that allows researchers to contemplate disease systems and derive sets of expectations regarding the behavior of these systems. The SIR framework (Keeling and Rohani 2008) provides a useful approach to presenting generalities of disease behavior in terms of the populations that are affected. For an evolutionary biologist, however, the SIR model and related ideas closely resemble the single-locus, haploid, closed-population models in population genetics: no one expects gene frequencies in any real-world population to behave as the models predict, but models provide excellent guides to what sorts of behavior can be expected. A comparison of these two types of heuristic models in two distinct fields is useful for another reason: one of the major challenges in population genetics has been that of taking the heterogeneity of real-world landscapes into account (Storfer et al. 2006). More or less realistic, spatially explicit, population-genetic models did not emerge until about a century after the origin of the field.

Epidemiologists have evolved a rich set of spatial statistical approaches with which to describe and delimit disease outbreaks (Pfeiffer et al. 2008). A field known as landscape (or spatial) epidemiology uses these diverse spatial tools to detect disease clusters, describe general geographic patterns of disease transmission, and test ideas regarding spatial pattern in case occurrences. Curiously, however, epidemiologists have not explored the idea of linking spatial patterns to environmental variation across the same landscapes in great detail. Many books in this field either do not mention environmental covariates at all (e.g., Lai et al. 2009), or discuss environmental factors only briefly in a single chapter near the end (e.g., Keeling and Rohani 2008; Lawson 2008). My interpretation of this situation is that epidemiologists often have not considered disease transmission as an ecological phenomenon that links multiple species and their respective interactions with the surrounding environments and, for that reason, have not pondered environmental covariates sufficiently.

In this chapter, I provide a few examples of space-only and space-and-environment models of disease transmission. The focus is primarily on reconstructing the geography of disease transmission risk, and secondarily on identifying specific factors that promote or reduce transmission rates across landscapes. Both real-world examples from the literature and hypothetical examples are explored.

Examples and Illustrations

As a first example, Joly et al. (2006) assessed the distribution of chronic wasting disease in deer across a portion of southern Wisconsin. They plotted out known cases of the disease; used a spatial scan approach to detect a single, major cluster; and envisioned the centroid of the significant cluster window as the "hypothetical introduction site," even though the precise centroid appears not to fall in an area of particularly high prevalence. They only marginally related this spatial cluster to environments, assessing the percentage of coverage by appropriate deer habitat as a covariate, but not considering additional environmental or geographic factors as potential modifiers to the basic spatial process.

As a richer example, Estrada-Peña et al. (2007) analyzed the massive outbreaks of Crimean-Congo hemorrhagic fever that began in 2003 in Turkey. They assessed the spatial distribution of known cases to detect spatial clusters and then used spatial scan analysis to test for areas with higher-than-expected numbers of cases at different spatial resolutions. They found such foci in 40 administrative districts when spatial scans were developed at the coarsest resolutions. They then went on to map habitat suitability for the tick vector based on remotely sensed data sets. They found that areas of higher disease case-occurrence reporting coincided with zones of high suitability for the tick, in tandem with interspersed agricultural, forest, and shrubby vegetation. This analysis thus took important initial steps toward understanding factors underlying the transmission of this disease, but it did not consider other potentially relevant environmental dimensions (e.g., climate, soil characteristics) that may eventually prove to be important in molding its transmission.

Space-only approaches to disease mapping will forever be vulnerable to several problems. First, they will depend on the geometry and the intensity of the sampling that underlies the patterns that are to be estimated. Second, if environmental gradients are anything other than smooth and broad, space-only approaches will miss any environment-driven patterns that are manifested on spatial extents finer than the sampling areas. Third, any attempts to transfer (predict) patterns from one area that is sampled to another area that has not been sampled will be difficult or impossible.

Contrasting the Two Types of Models

In general, space-only models present several disadvantages in comparison with models that include both spatial data and information from environmental variables that can educate the model predictions to varying degrees. For any zoonotic disease, and for many diseases with environmental causes, the pathogens are biological species that are subject to some configuration of the BAM diagram. It is clear that considerable information is available regarding environmental variation that can be marshaled to guide modeling efforts. In space-only models, this information is neglected; in space-and-environment models, environmental data are available to the models and may often enrich model predictions considerably. This difference is the explanation for the broad, simple patterns that are generally recovered in spatial analyses of disease transmission processes.

Moreover, the results of space-only models depend rather critically on the geometry of the positive sampling. If the sampling is anything other than completely random, or unless it is sufficiently comprehensive to have detected most or all cases, the spatial structure of the sampling can dramatically affect the outcome. Large spatial gaps in sampling or in the occurrence of diseases can lead to interpolation areas that, in space-only models, can easily produce artifactual patterns.

More generally, space-only models will be unable to "see" the fine details of a pattern unless those details are actually sampled. Environmental variation is structured hierarchically, with some dimensions that are manifested on continental extents (e.g., climates), and others that exist over smaller areas (e.g., land cover); species' spatial distributions respond to these drivers at multiple scales. Space-only models will be limited in any situation in which the sampling across space is other than complete and extremely detailed, and potentially even when the sampling *is* detailed, depending on how nonoccurrence information is managed in the analytical process.

One use to which these modeling approaches are commonly put is that of transferring models from one region or time period to another. If current distributional patterns present a certain scenario, then what can they tell us about likely future distributional patterns? Applications of this sort can include predictions over space to unsampled regions, anticipation of potential distributional areas under future climate conditions, and estimations of possible distributions in other regions. These model "transfers" are essentially impossible if the models are based on spatial patterns only, rather than on environmental drivers that can be estimated for both the region that has been sampled and the "other" area (another region, another time period, or another continent).

All of these problems will be amplified still more when occurrence data are not point based. As will be discussed in detail in chapter 9, much information is lost in this type of data: not only are any details smoothed over and hidden (e.g., Nakazawa et al. 2010), but differences in spatial extent across broader areas will cause further confusion, simply because greater detail is available in certain regions than in others (L. Eisen and Eisen 2007; Peterson 2008b). Hence any problems discussed in this chapter regarding space-only models will be dramatically magnified by the degeneration of occurrence information to polygon-based occurrence data, a common situation with disease-related data.

Models that incorporate environmental information, on the other hand, have the potential to remedy these problems, at least partially. If spatial sampling is incomplete or has gaps, but consistent associations with particular environments exist, it will often be possible to fill in the missing portions sensibly, based on the environments manifested there. Environmental data are available with considerable spatial and temporal details, which allow substantial specificity in model inferences, at least down to the spatial resolution of the input data. Therefore, in the remainder of this book my focus will be on models that seek to characterize occurrence-environment associations, and I will then use those associations to identify distributional areas that fit those environmental conditions. This type of exercise is what is best referred to as ecological niche modeling, since the occurrence-environment associations are the target of the model estimation process (Peterson et al. 2011).

Conclusions

This chapter has focused on differences between models that are based on spatial pattern only and those that also incorporate environmental variation. The clear conclusion is that more information can be brought into the equation when environmental variation is considered. In this sense, if the task at hand is to map disease transmission risk, the logical path to follow is that of incorporating environmental information into models whenever possible. Thus niche models have much to offer that space-only models do not.

Recent thinking, however, has found uses for both of these approaches (Waller et al. 2007). In a future generation of models, one can envision estimating the ecological niches of each component species in a given transmission system (pathogen, vector, host) to achieve the finest spatial detail possible. Space-only methods can then be incorporated into the process to characterize the spatial pattern of the filters and deviations—which do not have the underlying ecological niche as a causal factor—across broad regions. This melding of space-only and space-and-environment approaches may offer considerable insights in future applications.

PART III

PREPARING THE DATA

8

Garbage-In-Garbage-Out Principle

Ecological niche modeling is the focus of this book, which I see as a useful improvement and vital complement to the existing toolkit in spatial epidemiology. This chapter aims to plant in the mind of the reader the lack of magic in the process of calibrating ecological niche models, and the ubiquity of errors and biases that can mislead the process. Niche modeling is becoming enormously popular, but it is perhaps too easy simply to push a button and then interpret the results that come out. The process is considerably more complex, and the garbage-in-garbage-out principle holds quite firmly. If one is not extremely careful about the biases and errors that go into the model, those same biases and errors will be in the results of one's analysis, sometimes magnified considerably. Many of the topics treated here briefly are discussed in greater detail later in the volume, but it is important that they be contemplated throughout.

Few things that are easy to do end up being the best thing to do. In life, and particularly in mapping disease transmission risk, every decision has an associated cost. If one wishes to have truly precisely georeferenced occurrence data for a given species, it will be at the price of fewer available data. If one has very finely georeferenced occurrence data in hand, and wishes to develop a comparable fine-resolution map output, it will be at the expense of increased computing time. Every decision that one must make along the way in this process has a cost. This chapter speaks to data quality and its effects on model outputs; almost without exception, the price involved is time and hard work.

Whenever protocols and programs become well designed and widely accepted, it becomes easier to "run some models" quickly and see what comes out. Unfortunately, these quick-and-dirty efforts risk missing critical methodological details and allow biases in the input data to remain undetected. Those biases are transmitted throughout the modeling process and can affect what comes out in the results. This chapter is simply a reminder of the complexities involved and the need to assure data quality a priori, in order to avoid having "garbage" come out in the end.

Problems with Data Quality

Chapter 7 discussed the availability of high-quality occurrence data relevant to disease distributions and ecology and concluded that access to such data presents a major constraint on the development of robust, predictive models that can educate researchers about geographic patterns of transmission risk. The difficulties can be divided into two areas: random errors versus systematic errors (biases). The process by which the two categories produce problems in the outcomes is the same (a nonrandom representation of some environments in niche model inputs), since they basically constitute elements of the filter layer (figure 13, in chapter 6) offered by Waller et al. (2007). Their underlying natures contrast, however, so they will be discussed separately.

Occurrence Data: Random Errors

The models under consideration in this book are based on occurrence data that are related to environmental data sets. These data place a given "species" (a pathogen, vector, host, or disease case record) at a given place on Earth at a particular point in time. Different sources of such information will have fluctuating levels of reliability, since varying degrees of care will have been used in preparing, recording, editing, and verifying them. These considerations have important implications for the results of mapping exercises.

In a best-case situation, a knowledgeable researcher visits a multiplicity of sites and records occurrence data. Each data point is georeferenced with a GPS unit that offers a spatial precision of about 10 m. The researcher is an established expert on the identification of the species or disease that is being investigated, so she or he checks each record carefully for correct identification. If the mapping is to be carried out at a spatial resolution of 1 km, then we can assume that the data essentially have no positional errors. An uncertainty of 10 m on a 1,000 m grid translates into particularly low error rates, as only 3.9% of the points fall within 10 m of the edge of a grid square at random. Note that if the mapping were to be carried out at 30 m, then 88% of the points would fall within 10 m of the edge of a grid square at random, and the potential for problems would be considerably greater. With some degree of spatial autocorrelation (chapter 10), however, spatial errors of one or a few pixels are likely to be unimportant.

Most such analyses, however, are not best cases. Rather, researchers often must find data that were collected by others (probably nonexperts), frequently without the benefit of GPS technology for georeferencing, were intended for other purposes, and were subject to unknown and often poorly documented procedures for georeferencing and taxonomic determination. These numerous unknowns can produce considerable noise in "found" occurrence data; Jiménez-Valverde et al. (2010b) presented recent examples from the biodiversity world.

As an illustration of found-data complexities, consider 10,714 records of *Peromyscus maniculatus* obtained from the VertNet biodiversity data portal. Of this initially rather impressive data set, 20.1% of the entries were not georeferenced, 9.3% had unhelpful georeferences of 0° latitude and 0° longitude, 13.4% had no uncertainty information, and 1.0% listed an unhelpful and clearly incorrect uncertainty of 0 m. At the end of the winnowing process, only 6,010 of the original data records were useful for analysis: the distribution of uncertainty values (figure 20) indicates that only 16.1% of the data set would be usable at a resolution of 1 km, although this figure increases to 88.4% of the data set that would be usable at a 10 km resolution. As a consequence—merely considering the uncertainty in the georeferences—40% of the data records are not usable; perhaps this data loss rate is acceptable when one begins with more than 10,000 records, but many disease-relevant species are considerably less well known, and one quickly would see problems with insufficient data. It is worthy of note that other relevant questions also surround the spatial arrangement of the sampling, which would further reduce the utility and magnitude of the data available.

Identification errors form another set of random-type errors. In biodiversity data, such as information documenting occurrences of vectors and hosts, these errors can result from simple misidentifications, or from the taxonomy employed in a given database not matching that used in others, so the names do not coincide, even though they might refer to the same taxon. These errors effectively identify a species as something that it is not, thus introducing a further random error into modeling efforts.

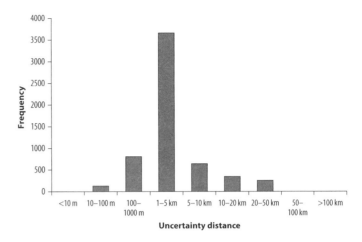

Figure 20. An illustration of likely amounts of spatial uncertainty in 10,934 records of *Peromyscus maniculatus* drawn from VertNet in May 2012. Of the initial total, 2,158 records were not georeferenced, and 999 had the meaningless georeference of 0° latitude and 0° longitude. A further 1,441 records had no uncertainty information, and 106 had zero uncertainty (which is impossible). Hence this figure is based on 6,010 records for which nonzero spatial uncertainty was reported.

In the most benign cases, these problems simply reduce the sample sizes available, while in more serious situations, they may introduce considerable noise and/or bias into the analyses.

In the disease world, different diagnoses may be accorded varying levels of confidence. For each disease, a set of diagnostics has generally been worked out, and these constitute what are called "levels of confidence" in the diagnosis. For Lassa fever cases in West Africa, for example, the highest level of confidence is ascribed when the diagnosis is via virus isolation, a polymerase chain reaction (PCR), an enzyme-linked immunosorbent assay (ELISA), or a plaque neutralization assay. Lower confidence is given to diagnoses via immunofluorescent assays (IFAs) or other laboratory tests, as well as to serosurvey results. Of 232 known putative Lassa fever cases, 28.0% fall into the high-confidence category, 63.4% into the intermediate category, and 8.6% presented no information whatsoever about the basis for the diagnosis (Peterson et al. 2014).

One might be tempted to assume that such random errors, at least if they are not overly common, will have no effect on risk maps, since they should only reduce the signal-to-noise ratio in the data. In a niche-modeling world, their effect is to place the species as apparently present under conditions that are, in reality, outside of its habitable possibilities. In this sense, the effect of

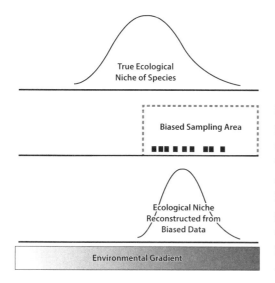

Figure 21. An illustration of how biased sampling across space can cripple the ecological niche modeling process. *Top panel*: a hypothetical true fundamental ecological niche of a species. *Middle panel*: sampling that is biased with respect to some environmental gradient. *Bottom panel*: the incorrect, biased estimate of the ecological niche that might result from an analysis of this situation. Adapted from Peterson 2005.

random errors is to broaden estimates of ecological niches and reduce the specificity of estimates of species' environmental requirements, although these effects can translate into significant differences in the configuration and extent of suitable areas in niche model outputs (Lash et al. 2012). This influence can be counteracted when error rates are contemplated and then included directly in the modeling and analysis process (Anderson et al. 2003), most effectively in terms of the E parameter (Peterson et al. 2008), to be discussed in greater detail in chapter 12.

Occurrence Data: Systematic Errors

The effects of random errors on risk mapping may be subtle, but the effects of systematic error sources can be much more prominent and obvious. Nonetheless, the latter are frequently neglected. Figure 21 shows an imaginary environmental gradient, with the species of interest having a niche that covers some subset of the gradient. If the sampling extends only over part of that gradient, the garbage-in-garbage-out principle indicates that any niche-modeling algorithm will recover the subset that was included in the biased sampling (Peterson 2005).

These problems can be avoided to some extent via careful delimitation of the areas of analysis. Warren et al. (2008) presented tests that referred niche comparisons to specific subsets (and their associated environments) of the overall study area; Barve et al. (2011) related these subsets to the **M** area in the BAM diagram. If we recognize that sampling may be limited with respect to **M** (not all of **M** may have been sampled well in terms of the environments presented), we can imagine another area—one that is well

sampled, so that any occurrences there are likely to have been detected—
and call it **S**. With this notation, we can now modify the arguments of
Barve and colleagues to focus on the part of **M** that has been sampled
thoroughly, or **M** ∩ **S**. This area is the region within which models should
most appropriately be calibrated.

In real life, these biases can be overt or somewhat covert. On the overt
side, one might be presented with monitoring data from a single state or
country. One can easily (with GIS) inspect the suite of environments associated with that region: to return to the notation presented in chapter 2,
this would be an assessment of $\eta(\mathbf{S})$. If $\eta(\mathbf{S})$ or $\eta(\mathbf{S} \cap \mathbf{M})$ is substantially
smaller and more reduced than $\eta(\mathbf{A})$, then a potential for subestimation of
the niche exists. This assessment is rather easy to achieve, at least if taken
one dimension at a time (e.g., figure 17, in chapter 6).

On the more subtle (covert) side, sampling and reporting biases can
still be present, even within a region that was sampled, and these biases
can affect the model's results. Spatial biases in biodiversity sampling are
nearly universal, with sampling localities clustered around sites that are
close to roads, near cities, known to have high diversity, have easy access,
and the like (Sastre and Lobo 2009). In a study of "roadside bias" across
Israel, Kadmon et al. (2004) found that biodiversity data suffer from this
bias in both geographic and environmental terms, which had significant
effects on the accuracy of model predictions, although these effects were
not dramatic. Thus the degree to which the environmental signature of
the sampled sites is not comprehensive will determine how compromised
the model results are.

Environmental Data Problems

Problems with environmental data used in the modeling process are much
less well studied than those with occurrence data, since it is usually much
easier to dig up more occurrence data than to produce a new global or
continental climate data layer. Investigators tend to assume that these data
layers are correct and not think too much about them. For example, the
paper presenting the WorldClim climate layer data set (Hijmans et al.
2005) has been cited more than 3,000 times, while researchers have rarely
developed their own, customized climate data sets.

Environmental data layers can have both random and systematic errors, just like the occurrence data. On the random side, erroneous data
from weather stations can be propagated through the interpolation process, producing anomalous estimates of risk areas. These effects will often
appear as peaks or valleys surrounding a single weather station, creating a
bull's-eye effect. Also, differences in the spatial resolution of various envi-

ronmental data layers can lead to mismatches and discords. These errors can easily be propagated through several series of analyses, and thus become nontrivial.

Still more worrisome are systematic errors (biases) in environmental data sets. A quite-common source of systematic errors is in climatic data sets, where densities (and data qualities) of weather stations can vary dramatically from one region to another, as can be appreciated in figure 22. These spatial patterns in the environmental data translate into differences in maps that are based on such data layers.

Biases Created by Geography

Perhaps the most pervasive biases are created by geographic realities. Consider figure 21, but, rather than using data from some biased sampling scheme, envision a biased representation of the environments on Earth, or one within the area that has been accessible to a species over its history ($\mathbf{M}$). The exact same effects that were discussed for figure 21 can result here: a truncated or biased estimate of the fundamental niche because of an incomplete representation of possible environments across these circumscribed areas.

Warren et al. (2008) presented ideas that were a fundamental step toward a solution; Barve et al. (2011) then translated their key concepts into the BAM framework. Anything that one might do in the world of ecological niche modeling must first be conditioned on the area that has been accessible to the species over its relevant history. To compare the ecological requirements of two species that have geographic distributions $\mathbf{G}_{A,1}$ and $\mathbf{G}_{A,2}$, respectively, one would derive the corresponding ecological niches $\eta(\mathbf{G}_{A,1})$ and $\eta(\mathbf{G}_{A,2})$. Any comparison of the two niches would not ask the simple question of whether $\eta(\mathbf{G}_{A,1}) = \eta(\mathbf{G}_{A,1})$, however. Bear in mind that these estimates generally underestimate the true niches, because the two $\eta(\mathbf{M})$s are neither comprehensive nor necessarily similar to each other, which forces the two $\eta(\mathbf{G}_A)$s to be different, even if the two $\mathbf{N}_F$'s are the same. The challenge would therefore be the tougher question of whether $\eta(\mathbf{G}_{A,1})|\mathbf{M}_1) = \eta(\mathbf{G}_{A,2}|\mathbf{M}_2)$, or (still worse) $\eta(\mathbf{G}_{A,1}|\mathbf{M}_1 \cap \mathbf{S}_2) = \eta(\mathbf{G}_{A,2}|\mathbf{M}_2 \cap \mathbf{S}_2)$. In other words, one would test whether the two niches are equivalently conditioned on the environments that have been accessible to each species *and* have been sampled.

Much confusion has been created by the use of niche-modeling approaches without a clear understanding of $\mathbf{M}$, as well as by the effects that this gap has had on model calibrations, evaluations, and comparisons. Barve et al. (2011) offered practical suggestions in estimating $\mathbf{M}$, which can be implemented with a few minutes' or a few hours' work, although the ideas

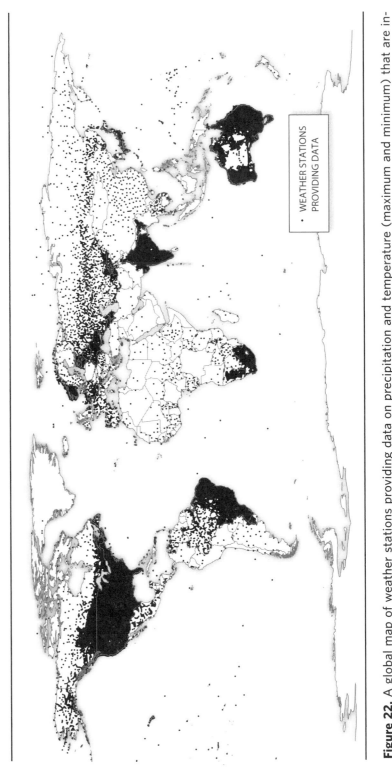

Figure 22. A global map of weather stations providing data on precipitation and temperature (maximum and minimum) that are incorporated into the WorldClim "bioclimatic" data sets. From the Global Historical Climate Network, ftp://ftp.ncdc.noaa.gov/pub/data /ghcn/daily/.

WEATHER STATIONS
PROVIDING DATA

they presented are far from final solutions. The most important first step is that the assumptions behind the choices of **M**, which define the area of analysis, be stated explicitly as part of the study methodology.

Conclusions

In the simplest sense, ecological niche models (and the transmission risk maps that can be derived from them) are based on nonrandom associations between occurrence patterns of species or disease cases and environmental characteristics. The algorithms by which such models are produced, however, are "dumb"—they are unable to distinguish between real biases in environmental dimensions that make up the ecological niche and artifactual biases that come from various data problems. This chapter has reviewed the sorts of errors and biases that may figure into the situation as a lead-in to a series of chapters that treat the details of how to develop ecological niche models. Although many of these topics will be touched on again in later chapters, it is quite important to keep these potential problems in mind from the outset.

9

Assembling Occurrence Data

One-half of the data input necessary for models from which the environmental requirements of species are estimated are the sites at which the species has been found to occur. These data are seemingly simple—just dots on maps—but numerous special considerations become important, in order to avoid problems that can flow through the chains of analysis and affect the final results. Moreover, occurrence data need to be compatible in time and space with the available environmental data, and they must mesh appropriately with the natural history and biogeography of the species in question. In the end, assembling occurrence data for a study generally turns out to take the most time and energy within the entire process described in this book.

Chapter 3 presented an overview of the availability of occurrence data and concluded that great imbalances exist. Biodiversity data are becoming well organized, allowing them to be censused and assessed. For a few vertebrate groups in some regions, huge amounts of data are available. For most other taxonomic groups and regions, however, data are partial or even lacking. Disease data are in a roughly parallel state, but without any overarching integration or organization. Hence investigators generally must assemble their own occurrence data; in doing so, they must be particularly careful not to introduce biases and errors into the data, and to detect and eliminate problems that are already present.

General Considerations

An initial set of considerations revolves around the precise nature of a data source. Occurrence data undoubtedly can be messy and difficult to manage. For instance, the idea of reading nineteenth-century specimen-data labels in handwritten script is not attractive to twenty-first-century researchers, who usually would prefer a nice, organized, digital-format summary. Such modern, digital, convenient data sets can have several deceptive features, however, that can exert insidious influences on model results.

Primary versus Secondary Data

Researchers initially must consider what the data records that are at hand actually represent. Primary occurrence data document a single occurrence of a species or phenomenon at one point in space and time. In this sense, such data records are unitary and serve to link a single occurrence of the species with a single set of environmental conditions. This type of data can be used in analyses at any temporal and spatial scale, and it can have multiple uses in quite different studies. For example, fine-resolution data for plague occurrences in Tanzania were the basis for a detailed study of landscape-level risk patterns across the Western Usambara Mountains region (Neerinckx et al. 2009a), and they were then able to be recycled for a much broader study of plague transmission across the entire continent of Africa (Neerinckx et al. 2009b). Primary data points are viable forever and can be used in any analysis for which they are compatible in terms of temporal range and spatial resolution.

Data of a more secondary nature are also frequently available. For mammals, detailed digital maps exist to document the geographic distributions of each species known from the Western Hemisphere. While it is seductively easy to use these data sets to summarize distributional patterns, it is also perilous: range boundaries in such polygon-based summaries are overly simple, particularly in light of scale differences, and the entire area within each polygon is not necessarily habitable for the species in question. Other types of secondary data products include atlases, range maps, and coarse-grid summaries of species' distributional patterns. These data products generally oversimplify a complex phenomenon or reduce a highly multiscalar phenomenon to a single scale. Hence, except in unusual or unavoidable cases, their use in disease transmission risk modeling is not recommended.

Polygons versus Points

In some situations—most frequently in the case of disease occurrence data—primary occurrence data exist only via references to polygons, such as a county, district, or state. These situations serve to locate the species in a place, but that place is crudely identified and specified. Although such data can be used in niche models (Peterson et al. 2004d; Nakazawa et al. 2007, 2010), since they offer geographic coordinates that can be input into the models, they nonetheless create significant problems and challenges in the resulting analyses.

The problem springs from the means by which the geographic coordinates represent the occurrence site. In initial attempts at analyses of this sort of data in a niche modeling framework, polygons were designated by the coordinates of their centroids, but this step introduced a base level of omission error that would then be present in model predictions and outputs (Peterson et al. 2004d). Newer studies have instead used multiple suites of random points within each polygon of occurrence (Peterson et al. 2006a; Nakazawa et al. 2007, 2010). These points more accurately represent the full breadth of environmental conditions across the polygon, rather than just the centroid, but the price is reduced resolution. Results of analyses based on such polygon-derived data tend to be overly general and lack much of the detail that is of intrinsic interest in such studies. This lost resolution most likely comes from false precision, because no confidence can be assigned to any particular point across the polygon's extent, and thus no true detail can be extracted from the data.

As was discussed in chapter 3, one major reform that can be made in the current system of recording, capturing, and archiving spatial information associated with disease occurrence concerns polygon-based georeferencing. This type of georeferencing involves an immediate loss of information (e.g., sometimes one knows the exact site of exposure, but at other times only that the person was in the western half of the county), but particulars are not included in the polygon-based georeferences. Also, since uncertainty measures are not used in combination with polygon referencing, crucial information regarding the precision of knowledge about the data points is lost as well. Although this situation is a fact of life for the present, the system of polygon-based georeferencing is clearly an idea would be best left in the past as soon as possible.

Real Spatial Resolution

Thus far in this book, I have used the term "spatial resolution" innumerable times, but without a careful operational definition. If I open Google Earth, it is reporting geographic coordinates to $10^{-6\circ}$, which in Kansas translates to about 8.6 cm. Are coordinates from Google Earth accurate to finer than 10 cm? If I am looking at the state of Kansas in Google Earth on my Macintosh laptop, the finest gradation (the difference between the coordinates of one pixel on the screen and the next one over) is about 0.007°, or roughly 660 m. The width of my cursor, however, ends up being about 0.2°, or about 17.2 km. So, what is the real spatial resolution of a data point that I locate with this system and then use that information to assign co-ordinates to a data point?

The lowest level of spatial resolution is set by the number of significant digits with which the coordinates are measured and reported; obviously, no more detail can exist than that. The underlying precision of the instrument used to create a georeference, however, must also be considered: whether it can distinguish between millimeters, or meters, or merely kilometers. Still worse is the extent of imprecision in the location of the item being recorded, either in the species under consideration or in that individual's exposure to the disease in question. In vertebrate scientific collecting, for reasons of expediency it is still customary to reference all animals that are caught to the coordinates of the central collecting camp, so the GPS coordinates of the camp may be accurate to 10 m, but in most cases the animal could have come from somewhere within 1–2 km of that site.

With disease cases, one has additional questions about where the person was exposed, or what the radius of possibilities was as to where that person was exposed. Depending on the particular disease system in question, in many instances, the subjects have no good idea as to where this exposure took place, but they can at least recount their activities over the relevant time window in which that exposure probably happened. The availability of such first-hand information is the reasoning behind the desire for point-radius georeferencing carried out at the time of diagnosis and data recording (an argument that was discussed in more detail in chapter 3). The question basically becomes one of uncertainty concerning the exposure site, given the movements and daily routine of the person in question over the critical time period within which exposure should have occurred. Hence the real spatial resolution of a data record—the uncertainty that should be assigned to it in its metadata—can be enormously crude and broad, but knowledge about how accurate the information is best represents how applicable it may be. This conflict between the spatial resolution of the observation (which can be quite coarse) versus the spatial grain of the phenomenon constitutes a significant and concrete limitation of the spatial resolution possible in these analyses.

Relevant Occurrences

A final caveat is that, with diseases that have different modes of transmission to humans or their associated animals, care must be taken to analyze only those cases that were infected by means that are relevant to the models being developed. Chagas disease is normally transmitted via bites from triatomine bug vectors, yet a fraction of cases that are contracted each year comes from blood transfusions (Ramsey et al. 2003b). Plague has a relatively rare pneumonic form that is able to be transmitted from human to human, rather than via flea vectors. Similarly, in a few instances filoviruses are transferred from their natural host to humans, but these viruses can then be passed from human to human in secondary infection chains. Distinguishing among these different routes that infections can take when assembling the occurrence data for a study is critical if the models are to be meaningful estimates of the environmental correlates of transmission, capturing, in essence, the ecological niche of the transmission phenomenon.

Obtaining and Improving Occurrence Data
Sources

Sources of occurrence data have already been reviewed to some degree in chapter 3, but here I provide a more detailed compendium of sources. For biodiversity data, the traditional and best-documented resources are systematic collections of scientific specimens (in natural history museums). Museum specimens have an enormous advantage over essentially all other data sources: they are identifiable and reidentifiable. As taxonomic concepts change and species limits are redefined, preserved specimens almost always can be reidentified to match the revised taxonomy. Although long maintained in analog formats, the data associated with museum specimens have been digitized in various efforts, beginning in the 1980s; more recently, major initiatives have integrated these data among institutions and now provide notably efficient access through Internet portals to distributed data resources (Peterson and Navarro-Sigüenza 2003; Stein and Wieczorek 2004; Guralnick et al. 2009). Nonetheless, in some cases no substitute exists for simply visiting museums and noting the data by hand.

For certain groups, however, the data associated with museum specimens are vastly outnumbered by observational data. Birds are easily observed, and in most cases can be identified readily if one has sufficient experience. Hence, in the case of birds, observational data offer the advantage of numbers, which can enhance certain types of research (e.g., Peterson et al. 2009). An ideal solution is to use both forms of information (when available), with the increased confidence in the specimen-based data linked to the greater numbers of the observational data.

A third category of data sources becomes more common with vectors, pathogens, and disease cases: project-based data sets, where previous or current studies have monitored distributions of one or more of these three biological actors. These data sets often are not well curated or documented with metadata, but they can be fundamental, particularly regarding the dynamics of vector populations in local areas and small regions. These data can allow researchers to address some of the more complicated components of disease transmission systems (e.g., dengue-transmitting mosquitoes, in Peterson et al. 2005), since specimen collections rarely provide that level of documentation and detail.

Georeferencing

Georeferencing is frequently necessary, since the original data often do not have a set of latitude-longitude coordinates associated with them, particularly in the case of historical or rescued data. Chapter 3 described the objectives of this process: (1) to represent the location of the site, (2) to document the precision of the location, (3) to document sources and methods, and (4) to retain all original data. That chapter also presented a summary of arguments regarding how best to achieve these goals. Here, I review these methods in greater detail; genuine, hands-on protocols and procedures are provided in several publications (Wieczorek et al. 2004; Chapman and Wieczorek 2006; Guralnick et al. 2006).

The first goal—that of assigning a best guess of geographic coordinates to a textual description of a place—is relatively straightforward. If the text describes a polygon (e.g., a state, a county, or a city that has a nontrivial footprint), then we assign the coordinates of the centroid (or multiple random representatives) of that polygon to the data record. If the text describes a point (e.g., a crossroads), the coordinates are obviously that point. Finally, if the text describes a location that is a set of directions from a point (e.g., 5 km north of Lawrence, Kansas), then the georeference must be displaced from the named point according to those directions.

The second goal is more complicated. Even in GPS-derived coordinates, some level of positional uncertainty exists (on the order of 10–100 m, depending on the circumstances). Textual locality descriptions, however, are much more uncertainty laden: places described as points actually have nontrivial footprints (e.g., the current extent of Lawrence, Kansas, is approximately 10 × 12 km), the measurement of "5 km" probably represents something that is not precisely 5,000,000 mm (perhaps a distance between 4.5 and 5.5 km?), and the direction "north" may not be exactly to the north (more likely a direction between northwest and northeast). As a result, the process of calculating overall uncertainty compounds all of these uncertainties (Wieczorek et al. 2004).

The remaining two items in the georeferencing challenge simply require good bookkeeping. The original data are never changed, so other investigators are able to return to that information and potentially reinterpret the georeference. The methods and information sources used to obtain the georeference are preserved in a set of data fields that carefully document how the georeference was derived, again permitting assessment of the interpreted information at a later date. The data and metadata fields must follow a well-structured and documented architecture, the best option being the Darwin Core standard (TDWG 2007b). The process is now partially automated, with a wealth of tools for obtaining the georeference (Guralnick et al. 2006; Rios and Bart 2008), calculating the uncertainty (Wieczorek 2001), and for the overall process of obtaining and documenting the georeference (Guralnick et al. 2006).

Data Quality, Error Flagging, and Data Cleaning

A further critical step is detecting, flagging, and correcting errors and inconsistencies in occurrence data. Data may hold internal inconsistencies, such as a country field that says "Brazil" but coordinates that fall in Paraguay; similarly, an older, more inclusive taxonomic determination may place a species (erroneously) at a locale that is outside its currently delimited range. At least these inconsistencies are detectable, if one is willing to compare information from different fields in the data record. Data records may also hold elements where information is not sufficient to reveal their inconsistency, thus promulgating the error. Chapman (2005a, 2005b) developed a set of best-practices manuals for data cleaning that provide additional detail on this theme.

Locational errors can be errors of detail, or sometimes ones of greater magnitude. Imagine a diagonal or irregular border between two countries, and relatively coarse coordinates documenting occurrence locations. The coarse coordinates can easily fall just over the border, creating inconsistencies between the country and the coordinates. These disjunctions make up the bulk of the problems in locational data. One important step in resolving this difficulty can be to inspect georeferences of occurrence data for records that have few significant digits (e.g., 4° N, 33° E). A small proportion of precise records will fall right on whole-number degree lines, but the great majority should have more detailed coordinates; such coarse-resolution coordinates should probably be removed from the analyses.

Taxonomic errors can spring from simple inertia and inconsistency, or from more overt misidentifications. A first step in cleaning taxonomic data is to compare them against authority lists: the nomenclature in a data set could be lined up against an accepted taxonomic treatment, and non-standard names could be discovered. Problems such as misspellings and errors of gender could also be detected by this means.

Taxonomic names have an odd evolution through time, however. One wide-ranging species (A) may be studied by a taxonomist and found to consist of two species; one of those two species will retain the original name (A), and the other will take a new or otherwise available name (B). This evolving set of nomenclature means that the name A corresponds to two markedly different concepts: one that includes the range of A and B, and the other that refers just to A. The only approach available to detect such problems is to check the data against current, authoritative references that summarize the geographic ranges of the species—a step that can be done manually but can be significantly time consuming if many species are involved.

Another means of detecting errors in either taxonomy or georeferences is to inspect outliers. This approach is highly effective, provided that the signal outweighs the noise in the data set (Chapman 2005a). One can plot points on maps, look for points at extremes, and give that subset of the data an extra check. In parallel, one can inspect how the points are scattered in environmental space and again check peripheral points in extra detail. In either case, some data points clearly have to fall on the periphery, but the idea is that any positional error may push points out from the main cloud, and these outliers should have a greater probability of containing errors.

For disease occurrences, one must consider the possibility of errors in diagnoses, which are akin to the taxonomic problems just discussed. In general, diagnostic techniques have improved steadily over the years, and newer test results should include fewer errors than older tests. Van de Groen et al. (1978) presented a very early study of the prevalence of antibodies indicative of Ebola virus infections among human populations using IFA tests. As expected, the authors found evidences of human infections in Central Africa; they also found a 6% prevalence in northern Rhodesia (= present-day Zambia), however, and a 4% prevalence in Panama. Hence some problems exist with building an in-depth understanding from tests that are so error prone, but methodology-based error rates can be addressed through comparative evaluations made using different techniques for each pathogen (e.g., Grolla et al. 2005).

An important step toward increasing the efficiency of these procedures to clean and improve occurrence data is by automating those aspects of the process that assess internal compatibility within occurrence data records. The dataCleaning module of speciesLink (http://splink.cria.org.br) is an excellent example of such a step. The tools in this module evaluate a data set in terms of its consistency in taxonomy, time frame, and georeferencing; suspect records are isolated and efficiently presented to the user for evaluation and correction (if warranted). This set of tools, however, is presently connected only to data sets in the speciesLink system, so its application to other data sets is not feasible.

Compatibility and Study Design

A final consideration in this chapter is that of compatibility among data streams. Compatibility between occurrence data and environmental data is quite important in the development of robust and accurate models. I focus here more on the occurrence data side, as they frequently appear to be the limiting quantity in niche-modeling exercises.

General Compatibility

In terms of temporal dimensions, compatibility is key. Niche-modeling algorithms relate known occurrences of species or diseases to their associated environmental context. If the occurrence data came from the 1950s, but the environmental data are from the 2000s, it may be that the occurrence was originally in a forest, but land use change now makes that site a cornfield in the environmental data set. At the very least, this temporal discord confuses the environmental signal in the niche model, and, in the worst situations, it may be seriously misleading.

Occurrence data and environmental data must also have compatible levels of spatial resolution. Formally, the finest spatial resolution for analyses is determined by the uncertainty in either data stream. The researcher can always filter occurrence data differentially, to permit higher levels of uncertainty and increase the sample sizes available, or lower the maximum acceptable uncertainty to improve spatial resolution in the analysis (but at a cost of reduced sample sizes). Hence this set of choices can and should be an iterative, experimental step in which different balances between minimizing uncertainty and maximizing spatial resolution and sample size are explored.

Area Sampled

Similarly, spatial components must be compatible. In chapter 8, I indicated that the study area should be determined by the accessible area **M** of the species in question, as this area is the one in which the species has "sampled" (has explored) over relevant time periods. Especially in an operational and empirical sense, **M** determines the outcome of basically all of the analyses and tests that are customary in niche-modeling exercises: model calibration, model evaluation, and model comparison. As such, assumptions regarding **M** must be made very carefully a priori and stated explicitly in all niche-modeling efforts (Barve et al. 2011).

M, however, must be modified if the entire extent of the area has not been sampled. Presence data do not need to document all sectors of **M**, but it is necessary for presences across the area to have some probability of being detected. Disease data, in particular, frequently will be sampled in relation to areas defined in terms of human activity: countries, states, or counties, for example. For this reason, the entire extent of an **M** defined biogeographically will frequently not be sampled. We have termed the area sampled by researchers for presences and absences as **S**. In this sense, then, the study area should be defined as the area **M ∩ S**, and all assumptions should be stated clearly in the methods section of any published results.

The strength of inferences regarding the dimensions and limits of the ecological niche of the species will be maximized as the calibration area is broadened, since the suite of conditions over which the model is relevant will also be improved (Elith et al. 2011). In this sense, although the definition of **M** is based on biological principles and therefore cannot be modified easily (Barve et al. 2011), every effort should be made to obtain a good, broad, comprehensive sampling from across the entirety of **M**, in order to have the broadest geographic area and the most inclusive set of environmental conditions represented in the comparisons that create the niche model. This step of maximizing inclusiveness may involve considerable extra work in pursuing relevant occurrence data sets that may offer information about distinct sectors of **M**.

Conclusions

The occurrence data that are used in developing models of a niche and predicting the species' potential distribution are a complex entity. They are apparently (but deceptively) simple in nature: just latitude and longitude coordinates of known occurrences of species. In reality, however, many considerations are important: primary versus secondary nature of data, point-based versus polygon-based qualities, internal consistency, uncertainty, and the like. Moreover, the area that was sampled to create the occurrence data set becomes a critical modification of the study area and determines the region that the study area should cover. In sum, these data must be assembled with great care, to avoid introducing new, and perhaps imperceptible, biases into the analysis and results.

10

Assembling Environmental Data

Referring back to the garbage-in-garbage-out principle (chapter 8), it is extremely important that the environmental data used as inputs in model calibration be assembled with great care and considerable thought. These data should summarize dimensions that are at least potentially relevant to limiting the species' geographic distribution. The choices involved are many and complicated, but they have significant implications for the success of any modeling exercise. This chapter reviews generalities relevant to these choices and specifics about the use of particular data sets in niche-modeling analyses.

This chapter examines considerations and important factors in choosing and assembling environmental data sets for niche-modeling analyses in disease mapping. The goal of this step is quite specific: to summarize relevant environmental landscapes over which the disease is or is not transmitted. The environmental variables included in this data set must be manifested across the geographic extent of the organism in question and must represent (or be closely and consistently correlated with) dimensions of the environment that are relevant (or potentially relevant) to limiting the distribution of that species.

More specifically, environmental dimensions in these analyses should ideally be those that participate in shaping **A** in the BAM diagram. It should be noted that, for practical purposes, the effects of interspecific interactions (**B**) may have consistent environmental correlates related to the ecological niche of the other species, but in the case of diseases, it may be difficult or impossible to separate the two. This intertwining of multiple ecologies is a reason why, to the extent possible, component-based approaches to analyses of disease transmission systems may prove to be vastly preferable to black-box approaches.

Relevance to Species' Distributions

Chapter 2 presented a detailed conceptual framework for thinking about and estimating distributions of species, centered on the BAM diagram. **A** is the suitable geographic area where the species can carry out its life history in full. The environmental combinations that make up **A**, referred to as $\eta(\mathbf{A})$, approximate the fundamental ecological niche $\mathbf{N}_F$, which is the object of the analyses that are being developed (see the cautions above regarding incomplete representation of environments across real-world landscapes, however). Hutchinson (1957) conceived of fundamental niches as being multivariate in nature, with the niche constituting a cloud of suitable conditions in a highly multivariate space of possibilities.

The goal of assembling environmental data is to put together digital raster data layers that mirror the dimensions of that relevant multivariate space as closely as possible. In some cases, the researcher may be able to find and include a data layer that summarizes one of those important dimensions, while in other cases she or he must settle for a data layer that is either causal to the desired layer (e.g., precipitation is related—and to some degree causal—to humidity) or is simply correlated with it predictably (e.g., elevation is closely and predictably correlated with temperature via the adiabatic lapse rate).

In practice, for most species and disease-related phenomena, little information exists regarding the identity of those critical limiting dimensions. For this reason, researchers generally initiate a study with a broad suite of environmental variables, or at least a broad suite of climatic variables. In some cases, investigators simply use all variables that are available (e.g., Fitzpatrick et al. 2007), which can cause significant complications, owing to problems with overparameterization and overfitting of niche models (Peterson and Nakazawa 2008). More frequently, however, researchers use procedures that either remove uninformative variables from their analyses or conserve the same original information, but in fewer dimensions (Peterson et al. 2011).

General Considerations
Interpolation and Information Content

One of the most vital aspects in choosing the environmental data sets to be used for analysis in niche modeling is the role of interpolation. It is important to understand how much of a data set is actually data versus how much is interpolation. Interpolated values will either be intermediate values between neighboring cells (a simple spatial interpolation) or will reflect the variation in some covariate: for example, elevation is frequently used to provide additional detail in interpolations of temperature across landscapes separating weather stations.

Moderate degrees of interpolation are probably acceptable, as even space-only interpolations may be sufficiently accurate for some uses. Few, however, realize and appreciate the degree of interpolation inherent in many of the data sets that are commonly used. The WorldClim data set (Hijmans et al. 2005) is based on analyses at a resolution of 30 arc-seconds (30″, or about 1 km^2) for weather station precipitation records from 47,554 locations, mean temperature records from 24,542 locations, and minimum and maximum temperature records from 14,835 locations (figure 22, in chapter 8). Globally, at a 30″ resolution, 933,120,000 grid cells are needed to cover the entire Earth. Relating the largest of the weather station sample sizes to the number of grid cells, only 0.0051% of the grid cells actually hold data; the remaining 99.9949% of the global picture is interpolated. The disparity between data and interpolation will be even more extreme for temperature-related variables. The coarser-resolution versions of this very popular data set are apparently derived from the fine-resolution original, so they still represent the same vast amounts of interpolation (Hijmans et al. 2005).

The issues surrounding interpolation are most clearly manifested in comparisons of ground-based data sets, which require measurements (e.g., weather station data), versus remotely sensed data sets, which generally have a unique measurement for each grid cell in the set. In the latter case, the images are made up of pixels, each of which is an independent measurement of some dimension of that portion of the overall landscape. Thus remotely sensed data may be much more informative than ground-based data sets; nonetheless, other considerations must be pondered, which are outlined below in the descriptions of individual data resources.

Additional important considerations are the characteristics of the data themselves, which could be categorical or numerical in nature. Although some modeling algorithms are configured to include categorical data in model calibration (e.g., Phillips et al. 2006), this type of data generally represents a considerable simplification of real-world environmental variations. Also, the sharp breaks that are inherent in categorical data frequently show up as sharp breaks in the potential distribution maps produced from them, which are unlikely to be biologically realistic. In the great majority of cases, therefore, categorical data are best not used.

Spatial Autocorrelation

A second critical feature of environmental data sets is the amount of detail versus spatial averaging that is inherent in the environmental dimension in question. Some environmental features, such as climate, tend to have broad and smooth trends over geographic regions; others, such as land cover, can be irregular and variable over even small areas. The tendency for values of a variable to be similar over distances is termed "spatial autocorrelation."

These differences in the amounts of detail in environmental variables are critical in reconstructing niches rigorously, for at least two reasons. First, an environmental variable that shows high spatial autocorrelation will not offer the model details at fine spatial resolutions. If one looks at changes in climatic variables across a small region, one tends to see broad, smooth curves that represent relatively little real information. To work at such fine resolutions and small spatial extents, one must seek other dimensions of the environment with less spatial autocorrelation, so that adequate detail is available.

Second, spatial autocorrelation is important in assessing the independence of the occurrence points available to document the range of a species. If one is analyzing an occurrence data set with respect to an environmental dimension that is highly autocorrelated over space, the points will not necessarily be independent of one another, and thus will not offer discrete characterizations of the environments used by the species in question. This problem can affect both the calibration and evaluation of the model: the number of points that are input into the modeling algorithm may not be representative of the true amount of information that is available to the algorithm; the models may be erroneous, due to an artificial overweighting of certain environments; and model significance may be overestimated because of inflated sample sizes.

Spatial autocorrelation is measured in the form of spatial lags. The study area is populated with random points, and these points are divided into pairs that are separated by different spatial distances (which have to be sorted into categories, or bins). Points that are immediately adjacent to one another will tend to have the same values for an environmental variable, while points that are far removed from each other will tend to have different values. The question is the rate at which these similarities decay over space, as can be appreciated in the comparison of different climatic variables shown in figure 23. Spatial lags are measured over each of the environmental data layers, and either the minimum or the median of these lags can be used as a spatial lag characteristic of a particular environmental data set.

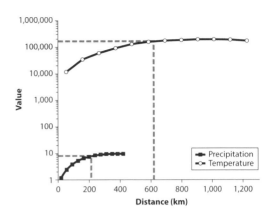

Figure 23. Spatial autocorrelation lag calculations, showing semivariograms for annual mean temperature and annual precipitation. In each case, dashed lines indicate the value for 80% of the spill and the corresponding spatial lag. The temperature data clearly have a broader spatial lag than the precipitation data. Note the $\log_{10}$ scale on the vertical axis, which compresses higher values.

In the strictest sense, these lag measurements should dictate which points can be used to train and evaluate models (although few studies have actually taken these rigorous steps). One could filter the available points so that no two are closer than the spatial lag of the environmental data set in question. This step—which would ensure that certain environments are not overrepresented because of clumped sampling points, rather than indicating the real preferences of the species being studied—is particularly important in model testing, where spatially autocorrelated data may inflate sample sizes and lead to Type I errors in statistical interpretation.

Dimensionality and Overfitting

The environmental spaces used in model development must be complex enough to capture the details of the ecological niche yet not overwhelm the usually limited quantities of occurrence data that are available for use in analyses. Analyses in highly dimensional environmental spaces too frequently will be overfitted: they will represent a close fit to the data on which they were calibrated, but their predictive ability will become poor when the model is presented with new, independent data. In niche modeling, the peril of overly dimensional spaces was clearly demonstrated in a series of studies claiming to have documented niche differentiation during species' invasions (Broennimann et al. 2007; Fitzpatrick et al. 2007). Among other problems, these researchers developed niche models in highly dimensional spaces, and for that reason were not able to use native-distribution-calibrated niche models to anticipate distributional potential in invaded areas (Peterson and Nakazawa 2008; Peterson 2011).

While it is important to ensure that the information content of environmental data sets is sufficient to permit the development of rich models, it is perhaps more important to avoid large numbers of dimensions. For some niche-modeling approaches, it is also critical to avoid highly collinear dimensions. These linked goals can be achieved in two ways: variable reduction or variable combination. In variable reduction, one calculates the correlations between different environmental variables over the study area (Jiménez-Valverde et al. 2009), seeks highly correlated pairs or sets of variables, and eliminates one or more of them, leaving a smaller set of relatively uncorrelated variables. In variable combination, one uses a principal components analysis or other factor-analysis approaches to seek linear combinations of original variables that express maximum amounts of differentiation while being orthogonal to one another. This second procedure can reduce the dimensionality of environmental spaces considerably (Peterson 2007b).

Tradeoffs between these two approaches exist, however. With variable reduction, researchers' decisions may not completely identify the most informative suite of variables, and information is clearly lost when some variables are removed from the data set. Nonetheless, these variables are easily interpretable to ecologically minded researchers. With variable-combination approaches, all variation is potentially retained, but the original variables (e.g., annual mean rainfall, minimum annual temperature) are bunched together into a complex function, and an easy and informative interpretation of the model's parameters is lost.

Spatial and Temporal Compatibility

A final general consideration is the compatibility of data sets in time and space (chapter 9). The extent of the data sets must cover the entire area of interest in the study, but the spatial resolution should not be finer than the uncertainty of the occurrence points to be used. If one finds discords in spatial resolution, either the environmental data should be coarsened (resampled) to become more compatible, or the filter for an acceptable degree of uncertainty (chapter 9) should be adjusted to match the environmental data set. Also, the occurrence data and the environmental data to be used in an analysis should cover the same time period. Otherwise, discords may enter the data set and confound the modeling process.

Modifiable Areal Unit Problem

A further consideration arises when data sets of differing spatial characteristics (e.g., grids with different origins and resolutions) are combined. Termed the Modifiable Areal Unit Problem (Openshaw 1984), this phenomenon is the result of changing grid positions and resolutions, which can produce different spatial statistical values. In some cases the problem may not be particularly serious (Guisan et al. 2007b), but in others it may be significant. The primary reason for pointing out this consideration is to make the user aware that combining data sets with wildly contrasting resolutions and positions is perilous. At a minimum, the methods used for any changes in resolution should be explicitly stated.

Specific Data Resources

The following is a short compendium of environmental data resources that can be considered for a given study. In each case, the researcher will have a series of tasks that revolve around standardizing projections and resolutions, particularly if multiple data types are used. The purpose of this step is to assemble a diverse and information-rich characterization of the environmental universe in which the species or disease phenomenon occurs. Ideally, the user would ponder the transmission system of the particular disease in question and what is known about its occurrence and tailor the choice of variables accordingly.

Climate Data

Climate data are clearly the most commonly used data resource in disease-related niche model development, especially in light of the easy availability of standard climate data sets (e.g., New et al. 1997; Hijmans et al. 2005), as well as in the development of suites of particularly informative "bioclimatic" variables. Climate data have important advantages. Their interpretation is quite straightforward, and much of the basic ecological/biogeographic theory was developed with these dimensions in mind (e.g., Grinnell 1917, 1924). Another important advantage is temporal range: for occurrence data sets that are older in provenance, frequently the only compatible data sources will be climate and topography.

Disadvantages, however, also exist. Climatic data generally have broad spatial lags. For this reason, extreme detail is not possible, as informative spatial resolutions would only be upward of 5 km or greater. Moreover, this amount of detail is generally not available, for lack of dense arrays of weather stations, thus demanding frequent and often ubiquitous interpolation of the data (figure 22, in chapter 8). Finally, because climatic data show considerable intercorrelation among variables, they are frequently collinear, so careful steps are required to reduce the dimensionality in climate-based data sets used for niche modeling.

Remotely Sensed Data

This data resource is perhaps the most promising and information rich for these analyses. Most of the remotely sensed (RS) data resources explored to date are based on characterizations across the electromagnetic spectrum of patterns of land-surface reflections. Because these raw measurements may be exceedingly complicated to compare and standardize, indices that are informative but also more stable in their variation are typically derived from them, such as the normalized difference vegetation index (NDVI), enhanced vegetation index (EVI), and the like (Bodbyl-Roels et al. 2011). Other RS products may characterize land-surface texture, vegetation structure, temperature, topography, cloud density, and other environmental dimensions, but these data resources have been explored far less in niche modeling.

RS data probably provide the richest single source of information that can be incorporated in such analyses. For example, the Moderate Resolution Imaging Spectroradiometer (MODIS) sensor on the National Aeronautics and Space Administration's (NOAA) Terra satellite offers bands of data characterizing 36 portions of the electromagnetic spectrum, with spatial resolutions of 250–500 m and daily coverage over most of the Earth's surface (Running et al. 1994; Justice et al. 1998; Huete et al. 2002). To remove artifacts of cloud cover and other complications, these data are generally assembled into 16–day composite data sets, and missing data pixels can be filled in by spatial or temporal interpolations. MODIS data are available for free from the Earth Observing System (EOS) data center, although the downloads can be a bit cumbersome, and a fair amount of data processing is usually necessary before they can be used in analyses.

Some MODIS composites provide a characterization of surface reflectance patterns across landscapes at a particular point in time and can be used to understand broad-scale patterns of land use and land cover. Much more informative, however, are multitemporal traces of the various indices through time. For vegetation indices, a high year-round index would be indicative of a continuously dense photosynthetic mass, such as a humid tropical rainforest; yearly patterns of high-and-low periods in vegetation indices convey seasonality resulting from changing temperature or moisture availability; and continuously low vegetation index values reflect desert or other nonvegetative land-cover types. Optimally, the user will take the time to assemble a large, multitemporal data set and process it into maximally useful and informative layers for analysis. One option is refining the data into variables that characterize seasonality and specific vegetation responses explicitly (Scharlemann et al. 2008); principal components analysis can be used to produce new, maximally informative environmental axes.

The biggest drawback is that RS data do not extend far back in time, since usable RS data sources did not come into being until the late 1970s and early 1980s, and the MODIS satellite was not launched until December 1999. To assure temporal compatibility, RS data cannot be used in tandem with older occurrence data. Another downside to RS data is that considerable work can be involved in importing and preparing these data sets for analysis.

Topographic Data

Topographic data are broadly available globally as digital elevation models, and they frequently offer additional information for niche models, in tandem with climate or RS data sets. The major topographic data sets include the Hydro-1K, ETOPO, GTOPO, and SRTM data sets, listed roughly in order of increasing spatial resolution and information content. Data on elevation per se are probably most frequently used in niche models, but they are often not appropriate. Elevation does not offer information regarding environmental conditions, but rather is largely a surrogate for temperature variation. It is a false surrogate, however, when broad latitudinal swaths are analyzed or when changing climates are considered. Nonetheless, other data products that can be derived from digital elevation models—such as slope, aspect, and topographic indices (the latter being an attempt to characterize a tendency to pool water)—can offer useful data dimensions with broad temporal applicability (since topography does not change quickly), fine spatial resolution, and some important environmental information.

Soils and Vegetation Structure Data

Until recently, data regarding soil and land-surface characteristics were largely categorical and typological in nature; such data will only rarely be useful and informative in niche models. More recently, however, several data sources that provide more quantitative characterization of soil characteristics have become available. The International Soil Reference and Information Centre's (ISRIC) World Soil Information section has recently produced a suite of soil maps for Africa at a spatial resolution of 1 km (ISRIC 2013); other efforts have produced a global map of vegetation canopy height based on Light Detection and Ranging (Lidar) imagery (Simard et al. 2011). These and similar data products have exciting potential to broaden the consideration of environments in the models described in this book.

Human-Related Data

In some situations the researcher may wish to add variables characterizing human presence and its impact in model development. These data sets are diverse and can include gridded human population data, indications of human economic development (e.g., nighttime lights), and vegetation disturbance. Because many of these data sets are derived from remote sensing data, they can often have quite-fine spatial resolutions but be limited in their temporal coverage, which can considerably constrain their applicability.

Conclusions

This chapter has summarized what is probably best thought of as drudge work in developing useful maps of disease transmission risk. The goal of the maps is to assemble a rich characterization of the environmental landscapes across which species are distributed, but a series of considerations (e.g., information content, relevancy, spatial resolution, temporal coverage, spatial autocorrelation, variable independence) must be pondered carefully. These data resources are increasingly available both conveniently and for free, and the researcher can select numerous data sets and dimensions creatively. It is generally possible to assemble a highly informative environmental data set for almost anywhere on Earth, so this task rarely limits or constrains analyses, at least for initial explorations. Nonetheless, the work involved may be considerable.

11

Study Areas and BAM

Assembling all of the input data correctly (the subject of chapters 9 and 10) constitutes almost—but not quite—all of the preparatory process for niche modeling. The points offered in this chapter, although they are recent lessons, are important in the development of robust and predictive end products. The delineation of the study area must be based on biological characteristics of the species under analysis and on the sampling available for that species. Moreover, the configuration of the BAM diagram for the situation under consideration and the relation of elements of the BAM diagram in environmental space become critically important. If the models are to be robust, the design of the study must be based on careful assumptions regarding the natural history, ecology, and biogeography of the species being investigated.

Until quite recently, most studies based on ecological niche modeling paid little attention to seemingly minor questions, such as the delineation of the study area. Projects arose that proposed to model everything via automated, stereotyped algorithms (Stockwell et al. 2006), where the development of a niche model was treated as a simple process, requiring little in the way of detailed parameterization. The reasoning behind these ideas—one tool with one parameterization was to be applied to any and all species—was decidedly against the tenor of the no-free-lunch theorem (Ho and Pepyne 2002). More recent efforts, however, have shown quite clearly that free lunches simply do not exist. Rather, each species and each geographic situation requires a more customized parameterization than automation can produce, at least with any ease.

This chapter treats these new insights into study design. The additional steps include (1) estimating **M** individually for each species involved prior to any analysis, (2) evaluating which areas within **M** have been sampled sufficiently, (3) guessing about the BAM configuration under analysis, and (4) evaluating the position of **A** with respect to **M** in the particular system being studied. These ideas run firmly in the face of any free lunches, since no single tool or single setup or single parameterization is likely to be robust in multiple situations. The earlier results are not necessarily invalidated, but they could certainly have been stronger. In new niche-modeling applications, however, it seems prudent to set up the studies appropriately.

Defining the Area **M**

I have already discussed **M** at several previous points in this book, particularly in chapter 2. **M** is the area that has been accessible to, and probably explored by, the species in the course of its relevant history (Soberón and Peterson 2005). Estimating this area can be complex and assumption laden; still, such estimates are quite critical in creating robust models.

Barve et al. (2011) pointed out the crucial role of assumptions regarding **M** in niche modeling. These authors showed that models calibrated under different assumptions about **M** arrive at markedly different end results, that the outcomes of model evaluations depend dramatically on which version of **M** was used, and that the conclusions from model comparisons (Warren et al. 2008) also depend on assumptions regarding **M**. In fact, **M** should be used to define the study area in model calibration in niche-modeling exercises, as it is the appropriate arena for comparisons. In spite of this overall importance, however, almost no publication has provided the biological reasoning behind the choices regarding its study area (= **M**) explicitly.

M can take on several manifestations, depending on the history of the species and the environment in the region being studied (Barve et al. 2011). If a species were newly arrived in a region, then **M** might simply be taken as the radius of individual dispersal distances around the known occurrence points. If the species had some history in the region, but not so long ago that environments had changed, then a certain amount of compounding regarding individual dispersal distances—perhaps several individual dispersal distances—might have to be included in the **M** estimate.

When the species has a history in the region that is long enough to cover periods of environmental change, however, then the effects of those changes on the species' past distribution should also be taken into account. This challenge is far from straightforward. Indeed, robust approaches have yet to be developed, although several suggestions and possibilities have been made.

1. Develop a simple niche model from random points distributed across the biome within which the species is distributed and project that model over the changes that the species has experienced (Peterson and Nyári 2007), thus providing a broad initial hypothesis of areas and environments that have been accessible to the species (A. Lira-Noriega, pers. comm.).
2. Use the distributions of other species sharing the same general range, as summarized in maps of biotic regions (Rojas-Soto et al. 2003).
3. Build simulations that consider niche and dispersal together (which, at least, are possible in theory). To date only "toy" implementations of such simulations have been developed (Barve et al. 2011), and such next-generation niche models have yet to be produced and tested.

Since methods for estimating **M** quantitatively are not yet mature and robust, a viable alternative is to make good, explicit assumptions based on a knowledge of the species and its biogeography. These assumptions should be stated explicitly and included as part of the methods section of any publication, just as one states α values for statistical tests and the reasoning behind Bayesian priors that are used. With clear information about assumptions, it is at least possible to evaluate the effects of particular **M** hypotheses on model results.

Sampling Considerations

As discussed at several points already, the study area should be established as the area that has been sampled both by the species (**M**) and by researchers looking for that species. Thus the area sampled (**S**) is used to reduce **M** to the area **M ∩ S**. The **M** part of this expression can be taken as the area that holds relevant presences and absences biogeographically: if the species is not there, its absence is not a result of limited access, but rather should generally be the result of unsuitable conditions in that locale. The **S** part of the expression indicates that researchers have been reasonably thorough in sampling for the species over the region, so that absences are not merely a consequence of insufficient sampling.

In some disease applications, delineating **S** is easy. Zendejas-Martínez et al. (2008) analyzed a detailed, government-developed data set treating bovine tuberculosis in the state of Jalisco, Mexico, in which cattle from the entire state were covered but neighboring states had no comparable information, so **S** could easily be established. Similarly, since plague is a reportable disease, in studies of plague transmission (e.g., Nakazawa et al. 2007), it was reasonable to assume that **S** covered the entirety of the United States, because any plague case anywhere in the country would be included in the data set.

In many other situations, however, delimiting **S** is much more difficult, as no obvious and clear sampling universe is necessarily defined. Presence records may be denser in some areas than in others, but it can be difficult to distinguish what is simply a region of abundant populations from what might be a region with exceptionally thorough sampling. On the other hand, a region might have no records because (1) the species did not find suitable conditions there; (2) it is present, but only at low densities; or (3) no sampling was done in that area. Hence it is difficult to distinguish biases that are part of the idea of ecological niches from biases that are artifactual results of uneven and incomplete sampling.

Anderson (2003) developed a framework for thinking about and procedures for establishing true absence of species, based on a characterization of the sampling background by using samples from other species that are sampled in similar ways to how the species of interest would be sampled. This reasoning—using other sets of species—could be applied fruitfully to estimating **S**. If the species of interest is collected or sampled in a way that samples other species as well, then one can potentially use the union of the samples of all of the other species as an effective estimate of **S**. Without such information, however, the only recourse is to make a best guess as to the area sampled, and state the assumption explicitly in the methods section; these ideas are related to use of a bias surface in some modeling algorithms (Elith et al. 2010).

BAM Configurations

The BAM diagram is a heuristic tool only, and as such should not be expected to be realistic or all encompassing. The general ecological and spatial relationships that it summarizes and simplifies, however, can be important frameworks for considering when particular analyses are appropriate and when they are not. The BAM configuration has substantial predictive power regarding the outcomes of modeling exercises.

The first inkling of this situation came in the downstream discussions from a high-profile paper (Beale et al. 2008) that indicated little or no climatic determination in distributions of bird species in Europe. The authors used an innovative technique that evaluated the cohesiveness and integrity of niche estimates based on known occurrences with those based on null models with similar spatial characteristics, in order to establish whether climatic factors were important. The authors, however, had failed to consider that Western Europe is surrounded by seas to the south, west, and north; to the east, the sampling available to the authors also ended abruptly. As a consequence, much of the distributional potential of the species they analyzed was limited by either **M** or **S**, rather than by **A**, because the very real climate constraints on those species' distributions might fall in the Arctic Ocean or the Sahara Desert. From the outset, then, the Beale et al. (2008) study was destined *not* to find any climate associations; a replicate study (Jiménez-Valverde et al. 2010a) carried out in North America, where **A** has a much greater potential to limit distributions, found near-ubiquitous climate determinations of species' distributional limits. Hence I refer to this error in reasoning as the "Beale Fallacy."

A recent, detailed analysis of these issues by Saupe et al. (2012) clarified the picture still further. The authors assumed that no significant constraints related to **B** were acting, and they studied the ability of various niche-modeling algorithms to reconstruct the niche under different configurations of how **A** and **M** might relate to one another. They developed a series of simulated (virtual) species for which niche and distributional characteristics were known precisely; they then sampled occurrence data from the distributions of the species and used a diverse set of niche-modeling algorithms to estimate those distributions. Under two of the scenarios ($A \approx M$ and $M \subset A$), niche-modeling algorithms did not perform any better than random expectations; significant, predictive models could be developed only when **A** and **M** partially overlapped or when $A \subset M$.

This body of work strongly suggests that an a priori assessment of likely BAM configurations should be an integral starting point in every study of this sort. The researcher must ponder the likely configurations of **B**, **A**, **M**, and **S**, based on the known distribution of the species, the configuration of environments, distributions of related species, and the like. In situations in which **M** or **S** is the dominant constraint on a species' distributional potential—and similarly for **B** under some circumstances—niche models are unlikely to succeed. Sometimes **S** can be broadened or spatial resolution can be sharpened, which can ameliorate the situation, allowing **A** to dominate $M \cap S$ in limiting that species' geographic distribution.

Details of **M** and **A** for Model Transfers

The situation is even more complex when the researcher wishes to project model results over a broader area or onto changing conditions. This procedure walks a tightrope between transferring model rules, which is simply the process of using fitted model rules to classify other sets of conditions, versus extrapolation, which extends model applications to sets of conditions that are not represented in the data sets on which the model was calibrated (Peterson et al. 2011). As any introductory statistics book will attest, extrapolation is perilous, as the model really has no information about what the response curve should look like outside of the conditions over which the model was calibrated.

Owens et al. (2013) clarified this situation in a study that used simulated species to explore the implications of different BAM topologies. Although occurrence data are customarily visualized in $\mathbf{G}$ (as dots on maps), niche models are fit in $\mathbf{E}$. Consequently, if one visualizes $\mathbf{M}$ in $\mathbf{E}$, whenever $\mathbf{A}$ approaches the periphery of $\mathbf{M}$, model rules can extrapolate or truncate response surfaces in ways that have little or no biological reality. Expressed formally, when visualized in $\mathbf{E}$, the set of habitable conditions $\eta(\mathbf{A})$ must be surrounded by a broad swath of unsuitable conditions for $\mathbf{A}$ within $\mathbf{M,}$ which can be denoted $\eta(\mathbf{A})^{\mathrm{C}}$, where the superscript C indicates "the complement of" or "not," rather than an exponent.

Since the goal of a niche-modeling exercise is to reconstruct $\mathbf{N}_{F}$, or at least $\eta(\mathbf{A})$, a problem arises: how to know the position of $\mathbf{A}$ if that is the goal of the study. In place of $\mathbf{A}$, the researcher must take the known occurrences (denoted $\mathbf{G}_{+}$), assume that generally $\mathbf{G}_{+} \subseteq \mathbf{A}$, and explore the topological relationships of $\eta(\mathbf{G}_{+})$ to $\eta(\mathbf{M})$. This sort of data exploration should be carried out before calibrating models. When the periphery is approached, any transfer of model rules beyond the limits of the area over which the model is calibrated, or to any other conditions (e.g., future climates), will run a considerable risk of wild extrapolation with little biological reality. All of the operations necessary to make these assessments are feasible within most GIS platforms.

Conclusions

This chapter has laid out four ways in which study design must be considered carefully in the contexts of the natural history, ecology, and biogeography of the species in question. The researcher must first ponder and estimate $\mathbf{M}$, next contemplate and estimate $\mathbf{S}$, then examine and guess at the likely configuration of the BAM diagram, and finally assess the degree to which suitable environments are surrounded by unsuitable environments within $\mathbf{M}$ when visualized in $\mathbf{E}$. None of these four steps is simple or easy, and each may involve guesswork and assumption. The critical point is that the researcher must be aware of the potential pitfalls and potholes in the modeling process and avoid them to every extent possible.

PART IV

DEVELOPING MODELS

12

Calibrating Niche Models

The central step in the ecological-biogeographic approach to mapping disease transmission risk that is outlined in this book is the challenge of estimating the fundamental ecological niche of each of the species involved in the transmission system under study. This process has been greatly facilitated by the development of user-friendly software platforms, but their ease of use must not outweigh the rigor of the biological context that the investigator imposes on the analysis. Biologically important aspects should not be skipped over, and this chapter presents a number of considerations that are central to calibrating ecological niche models appropriately.

As has been discussed in chapter 2, the fundamental ecological niche of a species is the set of conditions under which it is able to maintain populations, although actual occurrence is conditioned both on the species having access to the site via dispersal, and on the biotic environment being appropriate for that species. As was explored in chapter 6, the object of a disease transmission risk modeling effort can be either individual components of a disease transmission system (pathogen, vector, host), or the end result in terms of human or animal cases.

In this chapter, I assume that all of the preparations for modeling have been made: occurrence data are obtained and organized; environmental data are in hand and standardized; and the study area designed appropriately, with careful thought given to **M**, **S**, and BAM configurations, as well as the relative positions of **A** and **M** in environmental space. These topics were treated in part III and are necessary antecedents to implementing the ideas in the present chapter. Considerably more detail and background are provided in a recent book focused entirely on the topic of ecological niches and geographic distributions (Peterson et al. 2011), to which the reader is referred if a more in-depth explanation is required.

Introduction to Niche Models

A niche model is basically just a statement that one set of environmental conditions is more like the conditions under which the species is known to occur than another set of environmental conditions. Many means of estimating these models have been proposed, some very straightforward and others quite complex computationally (Elith et al. 2006), but they all distill down to the same challenge of identifying the subset of **E** that is suitable for the species.

A Simple Example

Perhaps the simplest of the niche-modeling algorithms is BIOCLIM, which is based on rectilinear climate envelopes enclosing known occurrence points. One identifies the range of values for each dimension in which the species occurs, from that derives the suitable range of values in a GIS, and then finds the intersection among the variables to identify areas that are deemed suitable. The result is easy to obtain but quite effective, and this approach has been used in many publications (Nix 1986).

As an example, I downloaded 82 occurrence points from the species-Link network for the bird species *Sclerurus scansor*, which is native to the Atlantic Forests of eastern Brazil. With a bit of quality control and the removal of duplicates, these points distilled into 19 unique occurrence localities, falling within an annual mean temperature range of 17.0°C–24.4°C and an annual precipitation range of 1,139–2,003 mm. If one believes that some error may be present in the occurrence data, one might remove extreme values. In this example, when 5% of the occurrence data that fell farthest from the median values was removed, the species' environmental range shrank to 18.7°C–22.4°C in temperature and 1,181–1,653 mm in precipitation. Plotting these areas on a map identified an almost entirely accurate range for the species in eastern Brazil as a hypothesis of $\mathbf{G}_A$, but the process also indicated areas along the Andean mountain chain, along the Gulf Coast and across Florida in the United States, in Central and East Africa, in Southeast Asia, and in eastern Australia (figure 24). In BAM terminology, these latter areas would be hypothesized as $\mathbf{G}_I$, or areas that

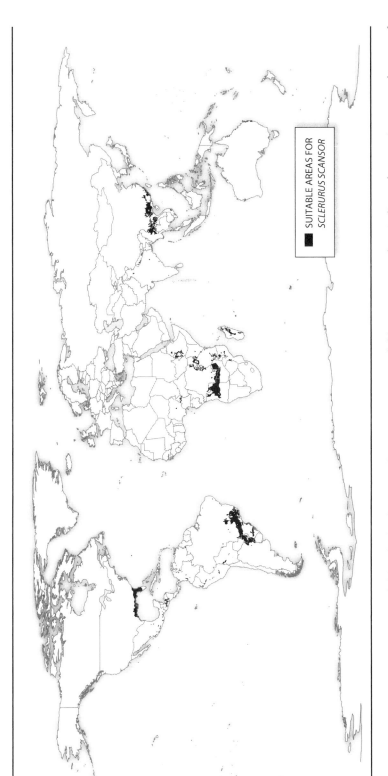

Figure 24. A simple BIOCLIM model for the bird taxon *Sclerurus scansor*. Available occurrence data indicated an environmental range of 18.7°C–22.4°C in temperature and 1,181–1,653 mm in precipitation for this species, which corresponds to the areas shaded in black.

SUITABLE AREAS FOR *SCLERURUS SCANSOR*

would be suitable for the species (in terms of these particular environmental parameters) if the species could disperse there.

Development of this niche model took a sum total of 15 minutes in a GIS program, and it could have been done even faster in the DIVA-GIS program, which has built-in modules for developing BIOCLIM models. The process is simple and straightforward, and it is a useful heuristic tool for exploring data initially. BIOCLIM models, however, suffer from various drawbacks: (1) model results are particularly dependent on sampling; (2) model predictions tend to become especially small when many dimensions are used; and (3) environmental dimensions are assumed to be uncorrelated within the modeled niche, so the models noticeably overpredict. Although BIOCLIM is no longer commonly used in publications, it is nonetheless a useful tool for thinking about and exploring data.

Niche-Modeling Algorithms

Two categories of data are input into niche-modeling algorithms: environmental data (chapter 11) and occurrence records (chapter 10). Different algorithms have widely varying requirements regarding absence information, which are often cryptic, as many algorithms obtain this information without it being supplied by the user. Hence, it is worthwhile to ponder these issues in greater depth.

Some algorithms require only presence information; the best example is BIOCLIM, which uses known occupied environmental ranges to construct niche estimates. Slightly more complex are various distance-based approaches, such as DOMAIN (Carpenter et al. 1993) and MinDist (Peterson et al. 2008), and a principal components–based approach published by Robertson et al. (2001). An index of suitability for all of them is based on the distances in environmental space to known occurrences. Ecological Niche Factor Analysis (Hirzel et al. 2002) has similar data requirements, being based on the position and breadth of the known occurrences of the species in a modified environmental space, although it does demand some information on conditions across the broader study area in order to fit factors.

A second level of information is required by a set of algorithms that portray known occurrences of species against a characterization of the entire study area. The principal and most-used example is Maxent, which fits a distribution of maximum entropy (a maximally spread-out probability distribution)—except for environments in which known occurrences force higher probabilities—and employs a user-determined level of smoothing over irregularities in the response surface (Phillips et al. 2006; Elith et al. 2011). Maxent thus needs information about the entire study area (background information), but it does not characterize absences. Other algorithms, most notably the Genetic Algorithm for Rule-Set Prediction (GARP), resample from the set of sites across the study area not known to hold the species as a proxy for absence data (such samples are known as "pseudoabsence data"), and then compare that information, in environmental terms, against known occurrences (Stockwell and Peters 1999).

Finally, a large suite of approaches requires absence data per se, including generalized linear models (GLMs), generalized additive models (GAMs), boosted regression trees (BRTs), multivariate adaptive regression splines (MARS), and random forests (Breiman 2001; Elith et al. 2006), among others. Although it is entirely feasible to generate pseudoabsence data and use them in place of true absence data (Elith et al. 2006; Phillips et al. 2009), this step violates many of the assumptions in these methods. In view of the uncertain nature of absence data in niche modeling, since many so-called absence sites will prove to hold either populations of the species or conditions suitable for the species to be present (chapter 10), the pseudoabsence data will not be fully representative of absence conditions, which causes considerable problems for these algorithms. For this reason, I focus my discussion in this book on the second category of methods—those using background or pseudoabsence data more comfortably—although it is also possible to use the strict presence-absence–based approaches in the broader framework described in this book.

Some of the diverse approaches in this second category merely separate the study area into two sets: suitable versus unsuitable sites. The best example of this class of approaches is BIOCLIM. At the other end of the spectrum, many presence-absence algorithms, as well as Maxent, purport to calculate formal probabilities of presence—although this assertion has been debated (Royle et al. 2012)—and Maxent is now seen as the equivalent of certain statistical models (Renner and Warton 2013). Nonetheless, probabilities of occurrence most likely cannot be estimated without rigorous comparisons of presence and absence data (Ward et al. 2009), while most niche-modeling approaches are best thought of as estimates of relative suitability only (Ferrier et al. 2002). Peterson et al. (2011) present a much more detailed treatment of these points.

Nuts and Bolts
Data Splitting

As the reader will appreciate from chapter 10, occurrence data are precious and hard won, but the imprimatur of thorough and rigorous testing, and the confidence that it accords to any sort of prediction, are equally desirable. For this reason, nontrivial proportions of available occurrence data frequently are sacrificed (i.e., not used to calibrate a model more rigorously) to permit a full and largely independent test of the model's predictive ability. This step will be treated in depth in chapter 14, but splitting up the data to make a model evaluation possible is a topic that at least must be mentioned at this point. Both model calibration and model evaluation need minimum sample sizes, although the data requirements for both can be quite low, if necessary (Pearson et al. 2007; Siqueira et al. 2009). More data is probably always better for both of these purposes, so sacrificing information will be a push and pull between the two sides.

The first question is how to split the data, which primarily depends on what is being tested. If the purpose of the modeling exercise is to anticipate transmission risk on the same landscape, then the most appropriate split will be randomized, although spatial autocorrelation (chapter 11) must be taken into account. A model based on a random subsample of perhaps half of the available data can be tested to see if it can anticipate the spatial distribution of the other half of the data (e.g., R. Williams et al. 2008), thus lending confidence in the model's ability to serve its purpose. On the other hand, if the purpose of the model is to predict into broad, unsampled regions, then it may be better for the data to be split spatially, so that a test of the model addresses this predictive ability.

Ideally, the model would be calibrated on one data set and tested on a second, independent one, so that biases within the data set are more likely to be revealed by the testing. An example is a study of triatomine bug distributions in Mexico (López-Cárdenas et al. 2005), in which existing museum-specimen locality data from outside the state that was the focus of the study were used to train the models, which were then tested using data from a new survey of triatomines across the state by health workers. Such rigorous testing, employing entirely different and independent occurrence data, lends considerable confidence to the assumptions that real niche parameters are being estimated, rather than biases and misleading patterns in the filters. Hence an important preparatory step is to pay particular attention to the best means of providing a rigorous model evaluation for a given situation, which will frequently involve splitting up the data at hand in some way.

Considering and Incorporating Error

A simple yet extremely important factor is what degree of error is expected in a data set versus what level of error should be considered "bad." Given that occurrence data are far from perfect (chapter 10), some errors in assigning environmental characteristics to occurrence points will undoubtedly occur. Most researchers have worked under the idea that more error is bad, and less error is good. If one can estimate how much error is expected, however, then reducing error too much may actually lead to overly inclusive and broad models, which is not desirable. In a treatment of model evaluation techniques, Peterson et al. (2008) suggested estimating a parameter E, which summarizes the proportion of occurrence data that is likely to hold meaningful errors that would change the environmental signal, either because of misidentifications or problems with georeferencing. This parameter is estimated via a careful exploration of the occurrence data, such as that which occurs in the process of data cleaning.

E also turns out to be quite useful in model calibration. In general, E provides a measure of how much error can be expected in a good model, so as not to penalize the model. Anderson et al. (2003) proposed generating large numbers of models by varying the input data (e.g., via the statistical practice of bootstrapping) and then sorting among them, based on measurements of omission and commission errors. Omission error is prioritized over commission error, because absence data frequently are not representative of a genuine lack of the requisite ecological niche conditions. E provides a useful tuning of this consensus approach: models presenting an error less than or equal to E are not penalized, while error exceeding E is counted against a particular model. The parameter E will also become useful—indeed critical—in model evaluation (chapter 14).

Variable Selection, Model Complexity, and Overfitting

Environmental spaces should not be overly dimensional, which leads to overfitting (described in chapters 10 and 11), yet they must be dimensional enough to inform models sufficiently. Typical strategies for managing this challenge have included the removal of redundant variables (e.g., Neerinckx et al. 2009b) and the use of principal components analysis to compact many dimensions into a smaller set of axes with less (or no) correlation among them (Peterson 2007b).

Model selection offers a new approach, as yet little explored but holding great promise. Although algorithms such as Maxent have internal constraints on model complexity (L1 regularization), approaches based on information criteria (e.g., the Akaike information criterion) can be used to set regularization parameters and achieve appropriate levels of complexity to maximize the information content of the model. This approach and related ideas will clearly see considerable exploration in coming years (Warren and Seifert 2011).

Calibrating the "Best" Model

A broad suite of papers has attempted to sort through the different algorithms and approaches available, in order to find the one that is theoretically "best" among them (Elith et al. 2006; Pearson et al. 2006; Guisan et al. 2007a; Ortega-Huerta and Peterson 2008; Elith and Graham 2009; Machado-Machado 2012). These studies typically set up some sort of predictive challenge, provided the same input data to different algorithms, and then documented which algorithms most successfully and most consistently anticipated the evaluation data. The winner was then taken as the best approach.

The publication by Elith et al. (2006) has now been cited more than 2,700 times and is clearly the champion among this genre of studies; thus it merits some exploration. The challenge it posed was to provide algorithms with presence-only occurrence data and see which could best anticipate patterns in independent sets of presence-absence data that were considered to be a gold standard. The study was massive, treating predictions from 16 algorithms for 226 species from 6 world regions. Hence these results have been taken as definitive, particularly in the authors' resounding endorsement of Maxent and a few presence-absence statistical approaches (e.g., BRT) as the best algorithms. Most of the huge number of citations to this paper in other studies refer to it to justify those publications' use of one of what it designates as the few best algorithms.

In reality, though, the no-free-lunch theorem (Ho and Pepyne 2002) suggests that no single solution exists for complex optimization procedures, such as estimating ecological niches (chapter 11). A nontrivial portion of the differences in quality between algorithms in the Elith et al. (2006) study has now been seen to be an artifact of the model evaluation methodologies they used (Peterson et al. 2008). A more balanced statement, cognizant of the no-free-lunch viewpoint, would be that a group of algorithms exists that has the potential for good performance. Some algorithms, such as BIOCLIM or distance-based approaches, may simply not be up to the complex multidimensional challenge of estimating niches. Promising algorithms include (at least potentially) GAM, MARS, BRT, Maxent, GARP, and neural networks.

Instead of choosing a single, supposedly best algorithm, a more productive approach would be to test two or three of these algorithms at the outset of a study, to see which appears to be performing better in that particular type of circumstance (e.g., species, region, environmental dimensions). Predictive challenges can be developed that are appropriate for the purpose of the study, the algorithms can be tested, and the one that is performing best in that specific situation can be identified. This algorithm can then be taken as the best choice for the challenge at hand, which would be much more consistent with a world that obeys the no-free-lunch theorem. An alternative approach, although not yet compared in detail with various schemas driven by the no-free-lunch theorem, is that of developing consensus models (Marmion et al. 2009).

Model calibration, then, is an exploratory process. Major preliminary steps include assembling the necessary and relevant data, controlling the quality of those data, and pondering and incorporating the biogeographic scenario. The researcher should then create some sort of test for the model predictions and assess a suite of potential algorithms for calibrating the models, with the best possibility for that particular challenge being chosen for model development, thus avoiding free-lunch thinking. At this point, the actual final model can be calibrated and evaluated, and its geographic implications can be explored and assessed.

Transferring and Extrapolating

Under some circumstances, the researcher will wish to calibrate models across one region ($\mathbf{M} \cap \mathbf{S}$) but project those model rules across a broader area: onto another region; or onto a separate set of coverages, perhaps representing conditions in a different time period. Peterson (2009) calibrated models for malaria across Africa under present-day conditions but projected model rules onto Africa that was characterized by modeled climates for 2050. Peterson et al. (2006b) calibrated models for *Cricetomys* rats (a possible host of monkeypox virus) on their native range in Africa but projected the models onto North America to assess the geographic potential of the species as an invader (since it is already established in the Florida Keys). In this sense, the model rules are projected onto data sets that may have different sets of environmental conditions than those manifested across the region where the model was calibrated.

Any statistics textbook will indicate that regression equations are valid only over the range of the independent variable represented in the data used to fit the equations; anything outside of that range is referred to as extrapolation and is not considered particularly valid. In niche modeling, model transfer occurs when the new area has conditions that are represented in the calibration area, while extrapolation happens when the new area presents novel conditions. The question is how to recognize these novel conditions, since environmental spaces can be many dimensional and extremely complex in their topology (Zurell et al. 2012).

Elith et al. (2011) proposed a method termed Multivariate Environmental Similarity Surfaces (MESS) that provides one means of identifying areas of extrapolation, and MESS calculations have now been incorporated into Maxent. This implementation is not completely appropriate, however, since it refers to environmental differences from the multivariate centroid of environments represented across the study area. The definitions of **M** and **S** summarize the area that the species has experienced and that has been sampled. Hence MESS should be referred to all of **M** ∩ **S**, rather than just to the centroid. Mobility-oriented parity (MOP), an approach developed in detail by Owens et al. (2013), is a modification of MESS-type methodologies. The point is that when a MOP-type analysis indicates that an environmental combination is outside of the environmental ranges of the calibration area, any projection of the model is extrapolative and therefore should be treated with considerable caution.

Characterizing Ecological Niches

If the purpose of niche modeling is to estimate an ecological niche, so predictions can be made, the environmental dimensions and limits forming that niche can also be described and delineated. These niche profiles, if interpreted correctly in the context of N_F^*, **M**, and N_F, can be enormously useful in understanding the environmental factors that shape species' geographic distributions. Multivariate statistical approaches, such as GLM, BRT, and others, have long been lauded because they offer an explicit characterization of the response surfaces of the model with respect to various environmental dimensions, and evolutionary-computing algorithms such as GARP have been criticized because they do not provide such information. These latter algorithms have been called black-box approaches (chapter 6).

The form of N_F^* and **M**, however, can be characterized quite conveniently, and more flexibly, via a posteriori analyses of the map output of any algorithm. This step is possible because the number of pixels in **G**, which is denoted |**G**|, generally corresponds uniquely to the number of environmental combinations |**E**|. Whenever the characterization of environments in the study area is reasonably detailed, a one-to-one correspondence exists between pixels and environments across the study area, or |**G**| = |**E**|. Given this relationship, one merely has to relate geographic grid cells (pixels) to the associated environments and the model's predictions. This step, achieved via some version of combining grids or extracting grid values to points, creates a single GIS layer with the values of all of the environmental grids and the corresponding prediction of the model. It is then easy to create diverse analyses and visualizations from this layer, as well as produce responses to values of the environmental dimensions in the model (Peterson et al. 2011).

Conclusions

Model calibration constitutes a critical stage in the risk mapping process, because—if it is carried out effectively—it is the step in which individual species' responses to environmental dimensions are captured. Many options exist for how to perform this calibration phase: not only are many algorithms now available, but many infrequently discussed but diverse options for data splitting, error management, and model transference also have to be pondered. The crucial point is to cast the modeling challenge in ecological and biogeographic contexts, so the model becomes best suited to the ecology of the species or phenomenon being studied. I have not reviewed here the specifics of the actual model-calibration methodologies here, as these steps are simultaneously well-documented and continually changing; rather, my focus is on key concepts and considerations to be contemplated in this process.

13

Processing Raw Outputs into Useful Maps

The challenges of niche model development do not end with successful model calibration, because the initial outputs of the models are no more than estimates of some quantity related either to $\mathbf{N}_F$ or $\mathbf{N}_F^*$, depending on the circumstances of the model's calibration. In most disease transmission risk situations, the researcher wishes to reconstruct a distributional area, but the simple operation $\eta^{-1}(\mathbf{N}_F)$, which identifies geographic areas corresponding to the fundamental ecological niche, does not identify the actual distributional area. Rather, in order to translate the model outputs into useful predictions, several assumptions are required about the biogeographic context of the species and its ecological niche.

Niche models that are calibrated following the ideas set out in chapter 12 produce either a map of pixels identified as suitable, or a suitability index, or (perhaps only in ideal cases) probabilities of occurrence. In one sense, this map corresponds to the abiotically suitable distributional area $\mathbf{A}$ or $\mathbf{G}_A$, but this area includes both of the areas $\mathbf{G}_O$ (the occupied portion) and $\mathbf{G}_I$ (the invadable area). Most studies about mapping disease transmission risk will wish to map $\mathbf{G}_O$, or at least will want to separate these two portions of the overall potential distributional area. Hence it is necessary to take a further set of steps—the subject of this chapter—that can transform the raw model output into a more usable product.

Choosing Appropriate Thresholds

Most algorithms for estimating ecological niches produce some continuous, or at least ordinal (0, 1, 2, 3, etc.), index of suitability across landscapes. Such views, however, can be misleading, since inordinate weight is accorded to the highest values, when lower values may be just as clearly within the conditions of the niche (Peterson et al. 2007c). As a consequence, in many cases a thresholding stage is required to convert the continuous values into a binary present-versus-absent view, and many methods have been proposed for achieving this step (Liu et al. 2005).

The nature of the presence and absence information in niche modeling must be used to guide the choice of a thresholding technique. Many (or most) thresholding approaches seek some type of balance between omission and commission errors (Liu et al. 2005); a much more appropriate method prioritizes omission errors over commission errors (chapter 3). If a taxonomic expert goes out in the field, collects a large number of samples, georeferences each site carefully with a GPS unit, and has given the correct identifications in light of her/his expertise, the researcher has ensured that the input data essentially have no errors; hence, every occurrence point should be within the final model. Regardless of the breadth of the area that is eventually identified as suitable (which may increase commission error rates considerably, depending on how large it is), all of the occurrence data should be included in the prediction. This approach thus prioritizes minimizing omission errors (leaving known occurrence points out) over any consideration of commission errors (including places where the species is thought to be absent).

The threshold just described has been termed the "least training presence" (LPT) method; in this book, however, I will refer to this approach as T_{100}, or the highest threshold that includes 100% of the known occurrences used to calibrate (train) the model. Returning to the E parameter described in chapter 12 (Peterson et al. 2008), which summarizes the proportion of occurrence data likely to hold environmentally meaningful errors, we can modify the T_{100} approach somewhat. In figure 25, occurrence data (circles) are clustered in areas of particularly good suitability, suggesting a very high threshold for establishing a prediction of potential presence. If the occurrence data include some errors (see arrow), however, then a T_{100} approach will push the threshold to a very low level, in effect identifying the entire area as suitable, which is not an ideal representation of the situation. With a nonzero estimate of E, one can allow some level of error to exist yet not have it influence the threshold. One would seek a

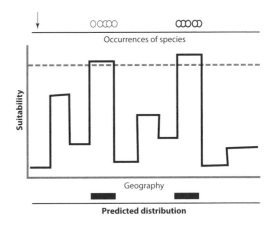

Figure 25. An illustration of ideas in omission-weighted approaches to thresholding continuous model outputs to produce binary predictions of presence versus absence. The figure covers a one-dimensional "geography" along the horizontal axis. Circles represent known occurrences used to calibrate the model. The initial view is that this species is only found in areas of high estimated suitability, and a least training presence approach would set a threshold at the dashed line that corresponds to a predicted distribution (the thick bars). If the occurrence data used in calibrating the model include errors (an example is the arrow), however, one would have to set a very low threshold and would recover an extremely broad predicted distribution. An approach that incorporates estimates of *E* (the proportion of occurrence data likely to include errors) would ignore the erroneous data point.

threshold that includes all occurrence data, but would be adjusted by E, which can be termed $\mathbf{T}_{100-E}$. If E is 0.05 (a common value), then one would be seeking the highest threshold that includes 95% of the training occurrence data. Thus that erroneous data point (the arrow in figure 25) could be ignored, which would result in a more accurate restricted estimate.

From Potential to Actual Distributions

The outputs from ecological niche models, even when thresholded properly, are still estimates of some type of distributional potential. These models were calibrated on the area defined by $\mathbf{M} \cap \mathbf{S}$ (the area that has been accessible to the species and has been sampled effectively), but the model rules can be transferred to broader areas. In particular, one would be interested in its predictions across $\mathbf{M}$, representing all of the suitable areas in this region (not just those portions that intersect $\mathbf{S}$) that should be inhabited by the species. In some applications, the researcher may be interested in the distribution of suitable areas outside of $\mathbf{M}$, which would indicate the invasive potential of the species. In other cases, even within the initial hypothesis of **M**, the researcher may wish to restrict the model predictions still further.

Refining **M**

M is an assumption set that emerges from BAM considerations (chapter 2), but estimating **M** is far from straightforward or easy (Barve et al. 2011), as has been discussed in chapter 11. Once niche model predictions are transferred to all of **M**, the dimensions of this initial assumption may become unsatisfactory. The researcher may define a relatively broad **M**, but niche model results may identify large areas of **M** as being suitable, but may also reveal that they were well sampled and never found to be inhabited. In cases like this one, a revision of **M** may be in order if the map is to reflect transmission risk accurately. Such modifications are a posteriori in nature, and should therefore be presented as lessons learned in the course of the research.

Considering Fragment Size or Isolation

In certain cases, the researcher may have some information a priori regarding minimal fragment size or the maximum degree of isolation that can sustain populations of the species of interest. In other words, some species would be known to have minimum area requirements, or could have maximum dispersal distances; either or both of these circumstances may constrain which patches of suitable conditions might actually be inhabited. These cases can be identified, often via software packets designed to assess fragmentation (e.g., McGarigal et al. 2002), and removed from the final model predictions.

Projecting and Transferring Models
Extrapolation

Chapter 12 discussed the perils of model extrapolation and suggested MOP-style metrics to assess the degree to which a given prediction (the suitability or unsuitability of a given site for a given species) is extrapolative. These considerations have been appreciated somewhat in disease mapping circles, such as the suggestion by Brooker et al. (2002) that "ecozones" be used to assess which model predictions are likely to be more reliable. Nonetheless, whenever model rules are projected to any conditions other than those on which the model was calibrated (**M** ∩ **S** at a given point in or period of time), caution is merited to avoid extrapolation and potentially misleading results.

In such applications, any model transfers should be accompanied by corresponding maps that summarize the degree of MESS-iness or some related metric (ideally MOP)—something that indicates how far outside of the calibration conditions (those represented within **M** ∩ **S**) each pixel in the broader area or other set of conditions really is. These maps of environmental distance thus summarize the degree to which one should accord confidence or uncertainty to predictions in a given area.

Dispersal

The role of dispersal shapes the next set of considerations, which are particularly applicable to situations where models are transferred broadly. In effect, in model calibration, one assumes that all sites within **M** are accessible to the species via dispersal. Although the initial assumptions regarding **M** can be revisited and revised, dispersal becomes a primary consideration outside of **M**. In this case, one is talking about shifting distributions of species either as invaders (if the species is able to overcome a dispersal barrier) or as a consequence of changing climates and the subsequent availability of new, suitable areas.

In past work in this field, either results have been cast as distributional potential, or extremely simple assumptions have been made about a potential dispersal of the species. For example, Peterson et al. (2001) presented three scenarios of distributional potential under different assumptions regarding dispersal ability: (1) universal dispersal, in which the species can detect and colonize any and all patches that are suitable; (2) contiguous dispersal, in which the species can only colonize suitable areas that share an edge with current distributional areas; and (3) no dispersal, in which the species can persist only in places where conditions do not become unsuitable. These scenarios are quite simple and generic, however, and a more mature approach would involve the direct simulation of dispersal processes.

A more promising approach explored by Barve et al. (2011) has yet to be implemented fully, but it will clearly be explored and developed in coming years. Under this concept, the niche and the species' dispersal ability are estimated at the same time, in the context of evolving environmental conditions. The idea is that the relationship of the niche to the evolving environment both creates distributional potential and molds the possible outcomes of dispersal attempts. These linked niche-and-dispersal models, which have been termed "third-generation niche models," will allow the incorporation of a greater number of realistic dispersal characteristics of species and will offer considerably better details regarding actual distributional shifts in response to changing conditions.

Conclusions

This chapter has presented a medley of modifications that may be necessary after the initial models are generated. Raw model outputs should generally be thresholded to reflect the primacy of presence (occurrence) data over absence data, assumptions about **M** should be revisited, assumptions regarding minimal patch size could be incorporated, and model projections to other regions or time periods can be examined in terms of their potential extrapolation and the possibility that the species can actually "get there" by dispersal. All of these modifications are designed to transform raw model outputs into an increasingly appropriate understanding of the actual distributions of species, or to produce clearer views of their evolving distributional potential.

Many of these steps require rather complex assumptions. As in any such situation, often the best approach is exploring various specific assumptions, to see how sensitive the results are to particular values. In this way, the researcher can begin to understand how the assumptions affect (or not) a given result.

14

Evaluating Niche Models

A model can be any prediction of an event or a possibility, and models are not necessarily good or informative or correct. Rather, the predictions from a model should be considered carefully only after it has passed rigorous evaluation procedures. This evaluation step is security against making interpretations from a model that has no predictive power regarding the phenomenon that is of interest in a study. Many recent publications present models that are not evaluated or that are not assessed rigorously. This chapter, instead of presenting an exhaustive review of possibilities, offers two robust options for carrying out effective model evaluations.

Model evaluation is generally a complex topic, one that merits a detailed treatment. In the field of public health, the need for rigorous model evaluations has been appreciated and emphasized (Brooker et al. 2002), yet many model results in this field and elsewhere are presented without any testing, or with testing that is not particularly thorough. In the interest of keeping this chapter to a manageable size, I refer the reader to a useful introduction to testing spatial predictions (Fielding and Bell 1997), and to a much more detailed and comprehensive treatment of model evaluation in ecological niche modeling in particular (Peterson et al. 2011). In this chapter, I distill the complex and convoluted overall situation down to two relatively straightforward solutions—one for binary predictions and one for more continuous predictions—that should yield rigorous and defendable model evaluations.

Controversies and Inappropriate Approaches

When models are tested, the methods employed have generally been what can be referred to as "softballs": simple tests that can easily be passed. A common practice is to split data into subsets for calibration and evaluation, using random subsampling. For models based on climatic data, which show broad spatial autocorrelation patterns, this procedure will frequently mean that the calibration and the testing data points are not independent of one another. This nonindependence is closely akin to using the same data point both to calibrate and to evaluate the model, which everyone would (or should) agree is circular, and not robust. Even the widely cited Elith et al. (2006) paper, which used distinct data sets for calibration and evaluation, did not control for a spatial separation of calibration and evaluation localities with respect to spatial autocorrelation lags, since they only checked that they had not used the same localities for both of these operations.

In my own work, I have long insisted on a spatial separation of calibration and evaluation areas, commonly in a checkerboard pattern. The justification has been to avoid autocorrelation effects on the independence of the points. Even with this well-intentioned procedure, however, autocorrelation effects on the independence of points *within* the evaluation area are not considered. Hence evaluation sample sizes may be inflated, since the points are in effect replicated artificially and are not necessarily fully independent of one another.

Evaluation statistics, as well as data-splitting techniques, have been controversial. Elith et al. (2006) used two evaluation statistics that, at the time, were quite novel in the field of distributional ecology: a correlation-based metric (that was not used subsequently), and the area under the curve (AUC) of the receiver operating characteristic (ROC). This latter approach has been widely adopted, as it has the advantage of testing across all possible thresholds, thereby avoiding assumptions about cutoffs between presence and absence. ROC AUC assesses predictive success (in terms of sensitivity, or avoiding omission errors), in relation to the proportional area identified as suitable across all possible thresholds, by tracing a curve of predictive success as a function of area. Performance and significance are evaluated by the elevation of this curve above a line that traces null expectations. ROC has now been implemented within the Maxent algorithm; as a consequence, it has been seen as the optimal approach to model evaluation and is used in many publications.

Lobo et al. (2008), however, pointed out a number of problems with the

ROC AUC approach: (1) it ignores actual predicted values and the model's goodness-of-fit, (2) it evaluates test performance over what are basically irrelevant predictions, (3) it weights omission and commission errors equally, and (4) it does not assess the spatial distribution of model errors; moreover, (5) the spatial extent over which models are evaluated (here, equivalent to **M**) turns out to determine evaluation outcomes (Barve et al. 2011). Peterson et al. (2008) echoed several of these points, particularly (2) and (3), and provided a solution to at least some of the failings of the ROC AUC approach that is detailed below. The primary point is that model evaluation methods have been weak and inappropriate, and they have not been seated in appropriate biological and geographic contexts. Hence few model results published in this field have been evaluated effectively and sufficiently.

Basic Concepts

The evaluation step for niche models can be a rather complex and intricate process. Therefore, this section explains some basic concepts that are necessary antecedents to the two evaluation techniques that are explained in subsequent sections. Each of these elements has important implications for the process.

Splitting Strategies and Hypotheses Testing

Each data set has intrinsic biases that paint a picture that is slightly different from the actual distribution of the species (or the relevant phenomenon) in the real world. Ideally, model calibration and evaluation would be based on independent data sets, with the two data sets collected by different means, which would thus be likely to have different sets of intrinsic biases (Peterson et al. 2011). If a model based on one data source is able to anticipate the distribution of points drawn from a different data source, then the effect of the biases has not been sufficient to cause substantial problems in model calibration. As an example, López-Cárdenas et al. (2005) used data drawn from existing museum specimens of triatomine vectors of Chagas disease to model the distributions of the species, and data provided by local health department surveys to evaluate the model's predictions. The independent provenance of these calibration and evaluation data inspires considerable confidence in the predictive abilities of their models.

In many or most cases, however, independent data sources are not available to researchers (one is generally happy even to find one data set). This situation requires splitting the available data into subsets: one for calibration and the other for evaluation. Several strategies have been used in splitting data: data can be subset randomly, or they can be subset by some spatial rule, such as a checkerboard or by political divisions. These different strategies, however, have important implications for the meaning of the model's evaluation and its outcome.

When data are subset randomly, the research question basically focuses on pure interpolation: the test data are interspersed with the model's calibration points, and the hypothesis being tested is the model's ability to anticipate the spatial distribution of a denser data set across the same landscape. On the other hand, when the data are subset spatially, the question can shift somewhat, focusing on the model's capability to predict across broader, unsampled areas. Hence various subsetting strategies for the data may correspond to different uses to which the models may be put.

Spatial Autocorrelation

An important modification of the usual protocols for model testing is to start by assessing the distances over which points are autocorrelated. Tobler's First Law of Geography—which underlies the idea of spatial autocorrelation—states that nearby objects will generally be more similar to one another than those that are far apart. In a niche-modeling universe, some environmental data sets are very broad and spatially averaged, while others are finer grained and less autocorrelated (figure 24, in chapter 12). Chapter 10 treated these topics when discussing the preparation of environmental data sets, but they also have important implications here.

Any points that are close to one another will generally not vary independently of one another, which has two critical implications for model evaluation. First, two points may end up with one of them in the calibration data set and the other in the evaluation data set, but these points may not be independent, which inserts a heavy dose of circularity into the evaluation process. Second, points that are too close to one another in the evaluation data set inflate sample sizes, to the extent that the statistical significance of the model's results may be expanded artificially.

If one is able to measure the spatial lag (the distance over which pairs of points will vary independently) associated with the environmental data sets (chapter 10), then these problems can be avoided. The overall data set can then be filtered to leave only those points that are separated by distances greater than the lag distance as part of the environmental characteristics being used to calibrate the model. With this modification (which can be rather severe in terms of reducing the sample sizes for model evaluation), points can be subset randomly without concerns about autocorrelation, or they can be subset spatially to focus the model's results more on predictions across broad, unsampled areas.

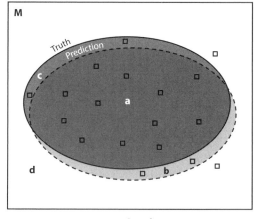

Figure 26. A summary of areas and data contributing to model evaluation. *Top panel*: geographic predictions (the gray area with a dashed outline) as compared with some truth (the gray area with a solid outline). Tests are carried out within the accessible area **M**, and evaluation data are shown as squares. *Bottom panel*: a confusion matrix. Letters denoting areas in the top panel correspond to those in the confusion matrix.

		Actual	
		Present	Absent
Prediction	Present	**a:** Correct Prediction of Presence	**b:** Incorrect Prediction of Presence
	Absent	**c:** Incorrect Prediction of Absence	**d:** Correct Prediction of Absence

The Confusion Matrix and Its Implications

Any binary prediction of a binary phenomenon creates a 2×2 matrix of outcomes: presences and absences predicted correctly and incorrectly (figure 26). This matrix, called the "confusion matrix" (Fielding and Bell 1997), has diagonal elements *a* and *d*, which represent correct predictions of presence (= presence of suitable conditions) and absence (= absence of suitable conditions), respectively. Element *b* summarizes incorrect predictions of absence when the species is really present; *b* is thus error of omission. Element *c* summarizes incorrect predictions of presence, when the species is actually absent; therefore, *c* is commission error.

In light of the discussions in chapters 3 and 9, the four elements in the confusion matrix must clearly be given different weights in any analysis related to niches and distributions. Among the correct predictions, *a* will be particularly important, while *d* will be artificial, since one rarely (or never) has rigorous data regarding absence of niche conditions. Error *b*, on the other hand, is a prediction of unsuitability in an area that in fact holds conditions suitable for the species. Element *c* is a model's assertion that an area is suitable, when in reality no data are available to the contrary. Element *c* should not be counted as true error in a variety of cases, because many locales are not recorded as having the species present, when in fact the species *is* there (figure 11, in chapter 3). This concept has been referred to as "apparent commission error" (Peterson et al. 2011), in light of its mixture of true error and simple interpolation and prediction differences. Proper management of this concept is crucial in distinguishing among appropriate and inappropriate approaches to model evaluation.

For instance, Cohen's kappa is frequently seen as a useful statistic in evaluating models. In kappa, however, elements b and c, as well as elements a and d, are weighted equally. This implies that omission and commission errors are of equal importance, which is rarely (if ever) true. Also, because b and d depend on data regarding absence, this information must generally be obtained via a sampling of pseudoabsences, which means that kappa will be sensitive to assumptions regarding **M** (Allouche et al. 2006). Similar balance problems in weighting error components bedevil traditional ROC AUC approaches (Lobo et al. 2008). In the remainder of this chapter, I avoid further discussion of any approach that equates the weights of b and c, as they are not applicable to the great majority of niche-modeling situations.

Binary Model Evaluation

Perhaps the simplest approach to model evaluation requires binary predictions. If the nature of the modeling algorithm does not produce yes-no predictions, then continuous model outputs must be thresholded (chapter 13). The areas identified by the model as suitable and unsuitable define the rows in the confusion matrix (figure 26).

The columns of the confusion matrix, however, must be drawn from real observations. Known presence data (occurrence data) provide the "actual present" column; the absence data are far more complicated, however, for reasons outlined in chapter 3. Because a genuine absence of suitable conditions can be established only under rare circumstances in a niche-modeling world, these data have generally been obtained via pseudoabsence sampling (Peterson et al. 2008). In such cases, it is especially difficult to decide what sample size is most appropriate for the absence data. With particularly large samples, pseudoabsence sampling becomes simply a characterization of the proportion of the study area that is predicted to be suitable, versus being unsuitable (Phillips et al. 2006).

In light of this one-sided confusion matrix, in which only presence data are actually based on information, a cumulative binomial test becomes the most appropriate approach to model evaluation. Generally, a thresholding step is necessary to produce a binary prediction. The area identified by the model as suitable, here termed $\widehat{\mathbf{G}}_A$ (the hat indicates that the quantity is only an estimate), covers a proportion of the study area $\mathbf{M} \cap \mathbf{S}$, which we will call p. If the known presence data ($\mathbf{G}_+$) are predicted with a number of successes (s) coming out of an overall sample size (n), then one can calculate the cumulative binomial probability of having at least s successes out of n attempts, when the underlying probability of obtaining an individual success is p. This probability is an indication of how unexpectedly well the known presences (evaluation data) coincide with the model predictions, while taking into account the null expectations of success.

The cumulative binomial test is powerful, especially in light of its simplicity. It uses presence data only, and merely tests whether the coincidence of those data with the model predictions is unexpectedly close. This approach requires the assumptions to be associated with thresholding model predictions (chapter 13), and it will be sensitive to assumptions regarding the extent of $\mathbf{M}$ and $\mathbf{S}$. This latter trait is characteristic of all such evaluation approaches (Barve et al. 2011), owing to the inclusion of broader or narrower areas with low probabilities of presence in conjunction with broader or narrower hypotheses of $\mathbf{M}$. A useful adaptation of cumulative binomial approaches to situations of small sample size was published by Pearson et al. (2007).

Continuous Model Evaluation

Evaluation techniques that do not rely on thresholds have been proposed as being greatly preferable to threshold-dependent approaches (Elith et al. 2006), although these advantages are less dramatic than originally portrayed (Lobo et al. 2008). Chief among the threshold-independent approaches has been the area under the curve (AUC) of the receiver operating characteristic (ROC). In this approach, one plots (1 – omission error) against commission errors across all possible thresholds, producing a convex curve that summarizes the accumulation of the model's predictive power as the anticipated suitable area is broadened (figure 27). Random expectations are the straight line connecting 0,0 (no predicted area, and 100% omission error) with 1,1 (the entire area predicted as suitable, and no omission error). The degree to which the observed curve is elevated above random expectations is summarized as the AUC, which must be well above 0.5 (the area under the random expectations line). Nonetheless, this approach has been criticized on a number of the fronts discussed above.

Peterson et al. (2008), however, modified the ROC AUC approach in several ways en route to a more appropriate methodology. First, following Phillips et al. (2006), they revised the x axis, so it no longer referred to commission error (which can only be estimated if absence data are available, with "absences" referring to the absence of suitable conditions, rather than simply to absence of the species). Peterson and colleagues used a

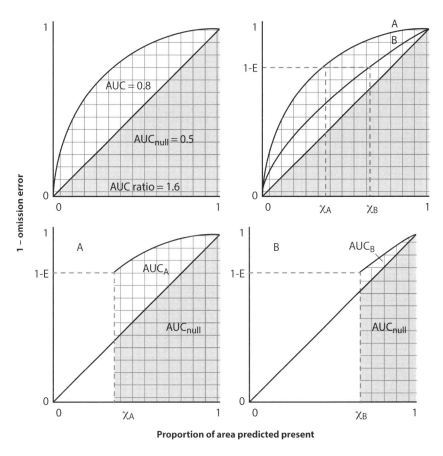

Figure 27. A summary of modifications to a receiver operating characteristic (ROC) analysis to reconcile it with key features of ecological niche modeling. *Upper left-hand panel*: a traditional ROC approach, comparing the area under the curve (AUC) of the test curve with 0.5, the AUC of the null expectation curve. *Upper right-hand panel*: a comparison of two curves, and an illustration of how the user-chosen error tolerance *E* identifies different critical area thresholds for the two curves. *Lower panels*: an illustration of the AUC comparisons that would be used to characterize each of the two curves. From Peterson et al. 2008, with permission from Elsevier.

simpler measure: the proportion of the study area identified as suitable at a given threshold. This step has the advantages of removing references to commission error and eliminating the model's dependence on absence data. Second, they restricted examination of the ROC curve to the portions that actually fell within the range of the model's predictions, because some algorithms characterize only portions of the full curve, creating artifactual differences in the AUC calculations. Finally, and perhaps most importantly, they limited consideration of the ROC curve to relevant predictions, which include all of the known occurrences of the species, subject only to the caveat that the E parameter (referred to at several points in this book) allows some nonzero omission error. Because the random expectations are no longer constant, the AUC is presented as a ratio of observed to expected areas under the curve.

This modified approach, which is a partial ROC implementation (Mc-Clish 1989), has the advantages of better balance for and weighting of different error types, appropriate management of absence data, and no overinterpretation of pseudoabsence data. It does, however, still require assumptions regarding the limits of **M** (Barve et al. 2011), as well as filters to account for the nonindependence of evaluation data because of spatial autocorrelation. Also, the method relies either on unthresholded predictions being fully informative (Peterson et al. 2007c includes examples of problems), or on the development of semicontinuous predictions from sets of thresholds (perhaps corresponding to assumptions regarding different values of E) that simplify the overall curve.

Model Evaluation and Model Performance

After a successful model evaluation, one can conclude that the model is indeed predicting the species' distribution (or other biological phenomenon) better than would be expected were the associations between the predictions and reality to occur only by chance. To achieve statistical significance, model predictions must be more accurate than correct predictions that could be expected through random chance. If we consider predicting the outcomes of coin tosses, random expectations are that 50% of the tosses will be successful; with a big enough sample size, a model that predicts this phenomenon with a success rate of 51% can be statistically significant, albeit not particularly useful or informative. Thus statistical significance is only part of the challenge; one must also assess the model's performance, to ensure that the model serves its intended purpose.

If the goal of a particular study is to identify all possible sites of transmission of a given pathogen, then the study will not be successful if omission rates are other than negligible. Alternatively, if the goal is to encounter a particular species that is especially rare or difficult to detect, then the commission error rates must not be too high. The goals of the study must be contemplated carefully, and indices and metrics (e.g., omission rates, correct classification rates, etc.) must be developed to indicate whether a given prediction will support those goals. Statistical significance measures should be paired with these performance measures in evaluating any model.

Conclusions

Model evaluation is complex, and it has led to considerable confusion, inappropriate or less-than-rigorous applications, and controversy in the literature. Initial assumptions and assessments are critical, such as getting the correct data-splitting design for the desired hypothesis, and evaluating the spatial autocorrelation structure of the environments. The tests themselves must be designed appropriately, so that portions of the confusion matrix are adequately weighted in model evaluations. Finally, if the product is to be genuinely useful for the application at hand, both statistical significance and model performance rates must be considered.

15

Developing Risk Maps

Up to this point, the book's focus has been on building effective and predictive models of distributions of species and other biological phenomena related to disease transmission. The presence of a functioning transmission cycle at a site or in a region, however, does not mean that human or animal cases of diseases will appear. Rather, several additional factors enter the picture, which offer additional challenges in transforming basic model outputs into a more realistic and genuine picture of transmission risk.

The distribution of disease transmission and disease emergence events is very uneven over the surface of the Earth. A recent analysis of 335 disease emergence events during 1940–2004 demonstrated nonrandom global patterns (Jones et al. 2008). These events were clustered in developed regions of the world, which were clearly a function of detection, diagnosis, and reporting biases. In reality, zoonotic and vector-borne emergences appear to be focused at lower latitudes, but the various filters and biases can obscure that view.

This book is about mapping the transmission risk of such diseases—disease by disease—across diverse regions, and it has addressed ideas of how to assemble a geographic view of disease transmission systems. In ideal cases, the researcher assembles geographic and environmental summaries of transmission systems, both in terms of the overall appearance of human cases and component by component, providing a picture of where

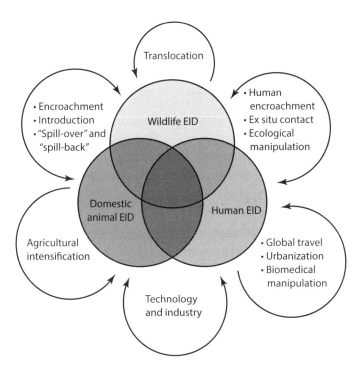

Figure 28. An illustration of the ecological continuum of parasites in which most emerging infectious diseases (EID) exist among wildlife, domestic animal, and human populations. Relatively few parasites affect one group exclusively. The arrows indicate some of the key factors driving disease emergence. From Daszak et al. 2000, © 2000, American Association for the Advancement of Science.

those systems should be manifested actively (where transmission occurs) and where they should not be functioning. Numerous additional factors modify the basic patterns of geographic occurrence, however, and these drivers must be taken into account if the maps that are produced genuinely anticipate patterns of transmission risk. These factors are the focus of the present chapter.

Daszak et al. (2000) discussed various factors associated with disease emergence events affecting wild animals, human-associated domestic animals, and humans (figure 28). They classified diseases that were related to free-living wild animals as (1) diseases of domestic animals that spill over to nearby wildlife populations; (2) diseases for which occurrence is a direct consequence of human intervention (e.g., via host or parasite translocations); and (3) wildlife diseases where neither humans nor domestic animals participate in the transmission cycle, but that occasionally spill into human realms. Their classification, and the complicated interactions pictured in figure 28, indicate how considerably complex this phenomenon is.

Initial Estimates

Chapters 11 and 13 have laid out a series of strategies through which a researcher can assemble initial estimates of the distributional potential of diseases. This process includes both black-box and component-based approaches, the latter comprising individual species-focused models for the pathogen (if it has a free-living phase), as well as for any host and vector species that are involved in the transmission system. This process must be accompanied by the full suite of steps with which to manage biases and confounding factors that have been reviewed in preceding chapters of this book.

These initial model outputs can be explored as preliminary, unfiltered views of transmission patterns, since they summarize either actual transmission patterns (black-box approaches) or a potential for transmission (models of key components, such as a competent vector). By a simultaneous estimation of black-box and component-based patterns, and a careful comparisons of the two, the spatial, environmental, and nonspatial filters for these patterns can be estimated and assessed. This approach, although not yet implemented in detail, is discussed and explored by Waller et al. (2007) and may offer considerable insight into disease transmission patterns. Regardless of whether these ideas can be implemented fully, the basic result that emerges from niche modeling must often be modified to reflect a series of additional considerations. These added factors range from biodiversity and species richness to human socioeconomic conditions and immune status.

Risk Modifiers

Disease transmission cycles may be manifested across diverse regions and landscapes, yet their transmission patterns to humans may differ from region to region. West Nile virus spread across the United States with massive numbers of human and equine cases, but the virus has behaved very differently in Latin America (Peterson et al. 2011). Similarly, human cases of infection with H5N1 avian influenza were numerous in Southeast Asia, but they have been scarce in Europe (Yee et al. 2009). The transmission systems are clearly functional in both cases, but some factors external to the actual transmission of these diseases appear to be modifying how humans and other associated species are affected (or not).

Biotic Dimensions

A new body of both empirical and theoretical work has outlined a series of effects that the biotic community has on disease transmission rates. These effects include the possibility of diluting transmission when alternative, less-susceptible hosts are present, thus constituting a series of blind alleys for pathogen transmission (Schmidt and Ostfeld 2001). Under other circumstances, disease transmission may be amplified by the diversity of the host community (Keesing et al. 2006). Nonetheless, the biotic community context is potentially of considerable relevance in determining transmission rates for a pathogen that is present in a region. These effects—at least in some systems—should be considered in assembling risk maps.

The work of Xavier et al. (2012) focused on Chagas disease control in the Amazon Basin. Spatially explicit analyses of the Chagas transmission cycle in wild mammals have been very few, so Xavier and colleagues related descriptors of mammal faunas to patterns of infection by *Trypanosoma cruzi* in dogs to identify hotspot areas of transmission across the Amazon region. The authors documented associations between the species richness of mammals and higher infection rates in dogs (figure 29). Several such studies of infection rates in relation to biodiversity have been published in recent years (e.g., Johnson and Thieltges 2010), with some confirming dilution, but others, amplification; hence the broader host community can be an important factor when developing models of disease transmission risk. Unfortunately, options for an easy characterization of host diversity levels are scarce, and this set of considerations will often require de novo sampling or modeling of large numbers of potential host species (e.g., Peterson et al. 2009).

Human Population

A more direct and tangible factor is human population levels. This element can function in a number of ways. In some cases, humans must be present, because they are a component in the transmission system (e.g., in malaria). Areas devoid of humans may have low transmission rates, and the entire cycle can (in theory at least) essentially grind to a halt. In other situations, incidental human infections may occur only when human populations reach particular densities, or diagnosis and reporting biases may be extreme in low-density regions.

Moffett et al. (2007), in mapping malaria risk across Africa, developed ecological niche models based on primary occurrence data for 10 mosquito vector species. They then incorporated the resulting maps into three relative risk models that assumed different ecological interactions

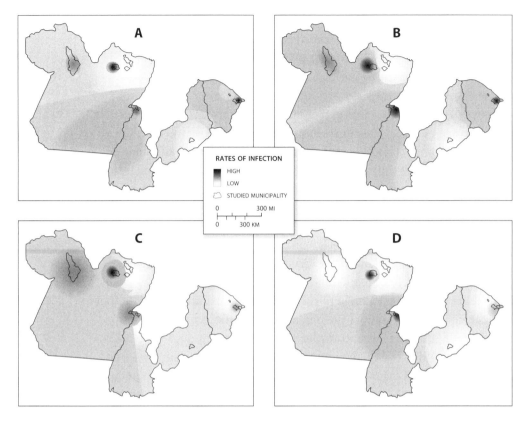

Figure 29. A visualization of associations between rates of infection with *Trypanosoma cruzi* in dogs for locales in the eastern Amazon Basin and northern Brazil, with (A) species richness of small wild mammals, (B) abundance of small wild mammals, (C) serological prevalence in small wild mammals, and (D) parasitological prevalence in small wild mammals. From Xavier et al. 2012.

between the vector species: "additive" models assumed no interaction; "minimax" models assumed maximum relative risk with any vector present; and "competitive exclusion" models assumed that relative risk accrued from the most suitable vector in a cell. The models in their study included variable levels of human association with the vectors, as well as spatial differences in human population density, in the development of relative risk maps. All of the models identified human population density as the critical factor in determining malaria risk across the continent.

Several human population data sets have been constructed, featuring reasonable degrees of detail across large regions of the Earth. The Land-Scan data set is the finest-resolution global population distribution data set available, attempting to present the ambient population (the average over a 24-hour period) for every 30″ grid square (about 1 km) on Earth

(LandScan 2006). LandScan uses complex algorithms to disaggregate census counts within polygons representing administrative boundaries across a fine-resolution grid. A very different alternative is the Nighttime Lights data set, extending back to 1992. This remotely sensed data resource shows light sources that are present during the night; it thus reflects certain types of human presence (those populations sufficiently advanced to leave electric lights burning at night), also with a spatial resolution of 30″ (NOAA 2012). These data sets (and others like them) have much to offer in terms of a rich characterization of where humans are to be found on Earth, featuring both global extents and fine spatial resolutions.

Other Human Dimensions

Numerous characteristics of human populations and their activities modify disease transmission patterns, and they should be taken into account in the process of mapping disease transmission risk. One example is land use and land cover (the actual condition of the land surface), which may not be included in niche models that are based on climatic data (particularly if occurrence data are older, and less precise). These landscape patterns and their changes have been implicated as strongly affecting transmission patterns of diverse diseases (D. Koch et al. 2007). Land cover patterns can be taken into account directly in niche model calibration only when the occurrence data are compatible in time and in spatial resolution with the land cover information (which may simply be a set of remotely sensed images that reflect land cover patterns). Otherwise, such data must be brought into the picture post hoc, after the models are calibrated when the investigators are in the process of assessing transmission risk.

Control programs and vaccination efforts are a second set of factors that have discernable effects in modifying transmission patterns over space. A region that has experienced intense mitigation efforts may not see much disease transmission, yet active transmission may occur if this area is immediately adjacent to another region where such efforts have been lacking. Masuoka et al. (2010) assessed the distribution of the primary vector species for Japanese encephalitis across the Republic of Korea, integrating mosquito occurrence data with raster data sets summarizing temperature, precipitation, elevation, land cover, and a normalized difference vegetation index. They found general agreement between the case distribution for Japanese encephalitis and suitability factors for the mosquitoes. Since much of the Korean population is vaccinated, however, Japanese encephalitis cases are rarer than one might expect, given the ubiquity of the mosquitoes. In addition, along the eastern coast of Korea, where pig farming is uncommon, transmission of this disease appears to be minimal.

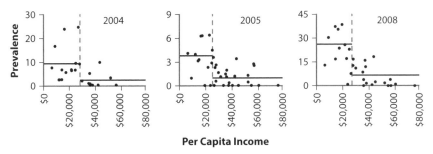

Figure 30. The relationship between average per capita income and West Nile virus (WNV) prevalence in mosquito vectors in Orange County, California, for 2004, 2005, and 2008. The dashed lines indicate the bifurcation between high and low prevalence values, as determined by tree regressions. The horizontal lines indicate mean values of prevalence for points above and below this bifurcation (Wilcoxon rank-sum tests for these means were significant for each year, p<0.001). Although absolute measures of WNV prevalence varied among years, relationships between predictors (per capita income in this case) and WNV prevalence were stable throughout the study period. From Harrigan et al. 2010.

Vector control programs can have macrogeographic effects in modifying disease transmission patterns. Chagas disease vectors (triatomine bugs) have been subjected to extensive control efforts, because the insects are relatively long lived. Human abodes that are fumigated tend to remain vector free, at least for a period of time, and transmission of the pathogen essentially ceases (Dias et al. 2002). These authors noted that in Brazil, a national campaign was initiated in 1983 to focus on the vector *Triatoma infestans*, with periodic visits to fumigate houses in affected areas. This species, which is exclusively domestic in the region, was successfully eradicated from practically the entire country. Many other triatomine species, however, maintain populations in adjacent natural habitats, so control programs attempted for them tend to work only for shorter periods of time (Ramsey et al. 2003a; Arias et al. 2012). Control efforts for these diseases and many others (e.g., malaria) can thus have major effects in modifying disease transmission risk patterns.

A third dimension of human landscapes that produces massive modifications in disease transmission patterns involves socioeconomic factors (characteristics of human populations that determine the conditions under which people live, and the places where they potentially contact pathogens). A recent study of West Nile virus transmission patterns (Harrigan et al. 2010) explored spatial transmission patterns in Orange County, California, a major transmission hotspot. The models explained 85%–95% of the variations in the prevalence of mosquito vectors and human infections.

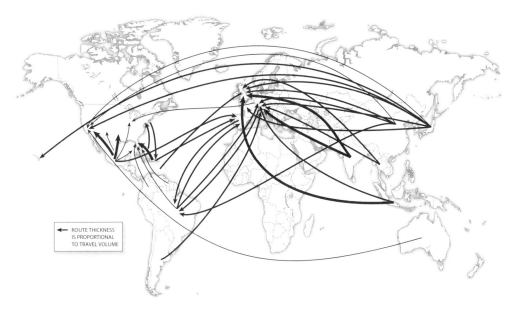

Figure 31. The 20 most traveled commercial airline routes entering the U.S. and the E.U. There are 40 total links, and the line thickness is proportional to the travel volume. From Gardner et al. 2012.

The prevalence measures they used were most closely related to economic variables (per capita income; figure 30) and anthropogenic environments (human population and neglected swimming pool density), illustrating the effects of features in the anthropogenic environment on the cycling and transmission of diseases.

A fourth important geographic element in human environments is that of human behavior, particularly movement patterns. A recent effort evaluated the risk of dengue introductions into North America and Europe, based on patterns of connectivity via people's movements (Gardner et al. 2012). Because humans are traveling on unprecedented spatial scales, the number of travel-acquired dengue cases has also been rising. The authors built statistical models on network information—basically, the connectivity of areas via international travel (figure 31)—and were able to identify risk patterns for these introductions. Movement and proximity end up being critical components in most models of epidemics, and any control of epidemics usually distills into reducing movement and increasing interpersonal distances.

Finally, a series of additional factors—characteristics of human populations or of host vertebrate populations—may modify disease transmission patterns. Preexisting immunity appears to drive the frequency and pattern of disease occurrence, such as for influenza (Andreasen et al. 1997). In other instances, immunity may depend on the genetic makeup of populations, as happens with malaria transmission and with the sickle-cell gene, which has uneven frequency patterns among human populations (Aidoo et al. 2002). The nutritional status of human populations clearly affects the infection dynamics of some diseases, as can coinfections with other pathogens (Badaró et al. 1986 contains a dual example). All of these elements may either modify local transmission patterns or have broader geographic implications, depending on the spatial pattern of the human drivers.

Type I versus Type II Errors

As with any predictive process that receives oversight and inspection from external users, risk mapping is a balance between widespread anticipation (to avoid missing potential positive sites) and specificity (to avoid crying wolf). While predicting broad areas may often be the best representation of true distributional extents, researchers must exercise effective management concerning overprediction and overgeneralization, in order to avoid false positives in model predictions and their associated lack of credibility (Pfeiffer et al. 2008). On the other hand, if models are overly specific or underpredicted, then disease cases may occur in locales outside the anticipated area, creating further problems with credibility.

An effective approach to these difficulties can be to provide models with a range of possibilities and scenarios. For example, Moffett et al. (2007) explored the implications of three relative risk models that assumed different ecological interactions between vector species; in the absence of solid information on which to base a decision among these possibilities, the exploration and presentation of all three models was a wise path to follow. Similarly, if precise estimates of the E parameter (chapter 12) are not available, one can explore low, medium, and high values, and thus bracket the likely true values. The idea is to hedge one's bets, supplying very specific information about areas most likely to have the highest risk, but also identifying broader areas of possible risk.

Overlay, Testing, and Simulation

By this point, one should have a sense of the transmission patterns of the pathogen of interest, as well as a set of possible modifiers to its transmission, infection, detection, diagnosis, and reporting patterns. Some means of combining and synthesizing this information is needed, to anticipate patterns emerging from the overall suite of data sets. Strategies for achieving this synthesis can fall into three major categories.

First, one can make clear and explicit assumptions about the transmission cycle and then overlay different data sets to create scenarios. For a vector-borne disease, one might combine maps of vector distributions and host distributions and assume that transmission will occur only within the area where they intersect (Peterson 2007a). These simple overlay approaches are quite sensitive to the quality and completeness of knowledge about the transmission system, but they can be clear and effective, at least in better-known cases.

A second approach is testing. One can adapt classic epidemiological approaches to assess whether particular data layers affect transmission patterns. Or, if one has a black-box summary of overall transmission patterns, as well as summaries of vector distributions, one can test whether the presence of each individual vector species is associated with an elevated risk of transmission. In each case, odds ratios can be developed and risk levels compared (after one accounts for spatial autocorrelation and the increase in apparent sample sizes that it can cause).

Finally, although I know of no comprehensive examples, simulation approaches can be used to explore these scenarios (Peterson et al. 2003 is a partial example). Instead of merely erecting assumptions and displaying their implications on a static map, each factor can be presented as a probabilistic effect, in which varying levels of risk are associated with different probabilities. If these methods are used, it is more feasible to incorporate uncertainty into the model outputs and present a more realistic summary of what is and is not known.

Conclusions

This chapter has described an amalgam of possible amending factors in shaping disease transmission risk across landscapes. The suite of these factors will be different for each pathogen and each transmission system, so this chapter has arrived at no firm conclusions and has offered little in the way of clear and concise guidance. Rather, it has presented a series of suggestions: knowledge must be assembled about the disease transmission system, relevant data sets must be sought out or assembled, scenarios should be explored and tested, and knowledge should be presented as clearly and as honestly (by portraying uncertainty as well as central tendencies) as possible. Only through this process can models ascribe the true dimensions (as best as are currently known) of disease transmission risk.

PART V

EXAMPLES OF APPLICATIONS

16

Identifying Risk Factors

The techniques explored in this book serve not only to map disease transmission risk, but also to examine the factors are associated with that risk, which can be useful for understanding and mitigating disease transmission. This sort of analysis has not appeared often in the context of the ecology and biogeography of the species involved. This chapter looks at the partial and largely cursory examples assembled to date and offers a vision for future studies.

Understanding the sources of risk is a central theme in epidemiology, and much energy has been expended in studies of risk factors. The literature is full of examples of research on outbreaks of a given disease, in which the relative risk of infection is assessed for a series of potential risk factors (e.g., Bausch et al. 2003). Although this body of work has illuminated many local transmission patterns, its focus primarily is on single sites or across circumscribed regions and thus may not be able to discern patterns and phenomena that are manifested across broader spatial extents.

With the ecological and geographic perspectives explored in this book, however, a broader viewpoint should be possible. This perspective might simply be an examination of which environmental factors are important in a black-box analysis of cases of a particular disease. A more in-depth study might assess environmental correlates of a key vector species' distributional ecology, thus addressing the role of an individual component of the transmission system. Finally, extensive calculations are possible, in which different factors are included (or excluded) from the model's development and the effects of their inclusion are assessed, but such approaches have not yet been applied.

Black-Box Disease Ecology

The simplest assessments of risk factors look for correlations between disease occurrence and environmental variation, in which all of the complex processes and interactions among species that lead to transmission are merged into a single measure of whether transmission is occurring. I call them black-box approaches, not pejoratively, but merely as an indication that the details of the process are merged and integrated, not considered individually. The strength of this approach is that it brings the entire, broad-scale view of variations in disease occurrence across diverse environments to light, linking and comparing them across many distinct factors. Environment-disease interactions that might not be perceptible in a study at a single site or across a small area may be much clearer when broader regions are analyzed.

Ellis et al. (2012) assessed a very broad suite of environmental variables to detect factors that might be important in the rangewide ecology of known monkeypox case occurrences. They evaluated patterns of known occurrences of the disease in relation to diverse environmental features, seeking variables that differed markedly between areas of known presence and areas without known occurrences. Their results identified annual precipitation, several temperature-related variables (primary productivity), evapotranspiration, soil moisture, and soil pH as factors of interest in shaping the distribution of the phenomenon. With this diverse and informative environmental perspective, the authors were able to determine that recent monkeypox cases in southern Sudan (in Damon et al. 2006) occurred under environmental circumstances that were quite distinct from all previously known cases. This conclusion, combined with phylogenetic information (in Formenty et al. 2010), allowed Ellis and colleagues to suggest that the Sudan cases were imported from farther to the southeast (the Congo basin), rather than representing an endemic strain. Similar black-box-based risk factor evaluations were developed by Donalisio and Peterson (2011) for hantavirus occurrences in southern Brazil.

A somewhat more in-depth assessment was made by Nakazawa et al. (2010), who analyzed ecological niches of different clades of tularemia to test the degree to which these various clades might (or might not) have distinct ecological niche footprints. Two subspecies of *Francisella tularensis* are recognized: *tularensis* (type A) and *holartica* (type B). Type A has been subdivided further into clades A1a, A1b, and A2, which differ geographically and clinically, but the distinctiveness of their respective ecological niches had not been assessed. In rangewide analyses by Nakazawa and colleagues, clades A and B had markedly distinct ecological niches, as did clades A1 and A2; clades A1a and A1b, however, displayed no appreciable ecological differences. The result was support for the idea that the categorizations of tularemia strains into clades A1, A2, and B have biological meaning (in this case, distinct ecological niches), but that the separation into clades A1a and A1b either is not a real biological division or at least has no ecological implications.

A more disappointing example involved an analysis of risk factors for dengue case occurrences in Medellín, Colombia. Arboleda et al. (2009) found clear ecological niche model predictions for areas associated with elevated dengue risk. The spatial predictions from the niche models were consistently corroborated by testing them with independent occurrence data sets. The study then assessed environmental correlates of these predictions, hoping to characterize the situations under which dengue transmission occurs in Medellín, but no single variable offered a distinct separation between high- and low-risk areas. A differentiation in suitability was manifested only in multivariate space, depending on complex combinations of variables.

Vector Ecology

One of the first studies using niche-modeling techniques focused on the Chagas disease vector species group *Triatoma brasiliensis* and the possibility of niche differences existing among the various forms that make up the species group (J. Costa et al. 2002). Ecological niche models were developed, based on occurrence data for each of four phenotypically differentiated *T. brasiliensis* populations. The model results indicated little transferability among the distributional characteristics of the different populations, suggesting that the four are ecologically distinguishable and that the complex consists of distinct populations at various points in the speciation process. These results, while published very early in the development of niche modeling and therefore not based on insights regarding the effects of **M** on model transfers (Barve et al. 2011; Owens et al. 2013), are still key in understanding the apparent differences in epidemiological characteristics among the four populations (J. Costa 1999).

Sweeney et al. (2006) assembled large data sets characterizing occurrences of mosquito species. They assessed environmental factors associated with the narrow coastal distribution of the malaria vector *Anopheles farauti* in Australia, using decision-tree approaches to examine the contributions by these variables, and then niche modeling to visualize patterns. These methods identified 7 specific climatic factors (4 related to temperature and

3 to moisture) from among the 30 environmental variables tested, which included 27 diverse bioclimatic parameters related to temperature, rainfall, and solar radiation, plus 3 variables summarizing aspects of landform and topography. Curiously, elevation also emerged as a crucial factor; if this information was not included, the models were unable to reconstruct the species' narrow limitation to coastal areas. The critical role of elevation here is probably related to the mosquitoes' need for estuarine waters, rather than fresh water. The results of this study illustrate the point that niche models can only "see" phenomena that are discernable and distinguishable in the environmental dimensions that are provided. The climatic variables did not differentiate among waters with different salinities, and the models were therefore blind to that dimension.

Human Variables

Early, but complex, analyses of the dynamics of dengue mosquito populations across Mexico indicated that the seasonality of environmental conditions explained a large portion of the distributions of *Aedes aegypti* across the country, and that these dynamics had predictive power regarding human case occurrences (Peterson et al. 2005). Yet those models neglected human-related variables (e.g., population, socioeconomic status, housing condition), and most likely were limited in their applicability and explanatory power. Machado-Machado (2012) analyzed the role of human variables in driving dengue occurrences across Mexico, based on human case distributions resolved spatially to the level of counties (*municipios*). At the spatial scale of the entire country, climatic variables proved to be more important determinants of suitability for dengue fever than the socioeconomic variables considered in her study. This result points to the dominant role of **A** in delimiting species' geographic distributions, superseding the roles of **B** (in this case, interactions with humans as another species). These results, however, most likely are highly scale-dependent; at some (finer) resolution, such human factors become quite relevant.

A study by Harrigan et al. (2010) on West Nile virus in California painted a very different picture, probably reflecting the finer scale of their analysis (chapter 15). West Nile virus prevalence in both vectors and humans in Orange County, California, was best explained by economic variables and anthropogenic characteristics of the environment. These results are perhaps not surprising. At some finer spatial scales, climatic variables (which are highly averaged across space and through time) should become irrelevant, because of their broad spatial autocorrelation structure, and more local-scale variables can take over.

Improvements and Future Steps

The ideas and examples presented in this chapter are partial and preliminary. In no case is a clear and detailed analysis available that crosses all the relevant scales and resolutions. Rather, the reader is left with tidbits and suggestive indications. It is to be hoped that, as this field develops further, more and better examples will emerge.

This set of applications illustrating ecological and biogeographic approaches to disease transmission systems, however, still has much to offer. If one could develop both component-based and black-box views of the same disease system, it would then be possible to make geographically educated calculations of relative risk. Comparing the two views, one could assess areas containing and areas lacking particular components (e.g., a vector species) and estimate odds ratios (Cromley and McLafferty 2002) that would be associated with those specific components. These calculations, particularly in comparison with others measured on more local extents, would provide a rangewide perspective that is presently completely inaccessible.

17

Spatial Interpolation and Prediction

The central theme of this book is mapping disease transmission risk, which involves interpolating among sites where a disease is known to be transmitted and projecting or predicting its movement to new sites, in order to evaluate its transmission potential there. This task is one of spatial prediction, but in the case of niche models, the predictions are based on environmental associations. These predictions can frequently be tested by calibrating the models on one data set and transferring model rules to another region where independent data are available. Once evaluated rigorously, the predictions can be used as risk maps or as elements of risk maps.

Perhaps the greatest strength of the techniques explored in this book is the idea of predicting distributional patterns over space (chapter 7). These methods contrast with most approaches currently in use in spatial epidemiology, since they take advantage of a broad array of information from environmental covariates, such as data on climatic variation or land cover characteristics. Subject to the caveats that the phenomenon must have an ecological niche and that the researcher must be able to summarize the dimensions of that niche in geographic terms, this additional information allows inferences across areas that have not been sufficiently sampled to detect foci, as well as provides considerable extra detail that is not available with purely spatial approaches.

Models can be good or not so good. As has been emphasized throughout this book, the process must be founded on solid thinking about ecology and biogeography, input data must be carefully quality controlled, and
models must then be calibrated and evaluated appropriately. Raw niche
model results must be thresholded and postprocessed, and only then
should risk maps be developed. These steps are the sum of what has been
discussed in chapters 9–15. The end result should be a spatial interpolation that predicts occurrences of species and biological phenomena (e.g.,
disease transmission), with considerable detail provided by the environmental covariates. This emphasis on prediction, however, may often come
at the cost of less-explicit explanations of model drivers.

Black-Box Examples

As was discussed in detail in chapter 6, the above techniques can be applied
either to components of disease transmission systems or to the end results
(occurrences in humans or in animals of interest). Classic examples of the
latter, which are black-box models, include a series of papers treating the
ecological niche of avian influenza cases across Asia, Europe, and Africa:
detailed analyses were developed for Southeast Asia (Gilbert et al. 2008),
West Africa (R. Williams et al. 2008), North Africa and the Middle East
(R. Williams and Peterson 2009), India (Adhikari et al. 2009), and Europe
(Si et al. 2010; R. Williams et al. 2011). In each instance, no clear information was available indicating how the input occurrence data related to the
disease's transmission to humans or poultry; human, poultry, and wild
bird cases were generally mixed together in the input data. Hence these
were truly black box studies, since little in the way of a mechanism or
individual components could be parsed out and analyzed independently.

A more fine-resolution analysis, and perhaps a more controlled one,
was that developed by Neerinckx et al. (2009a) for plague occurrences in the
Western Usambara Mountains of Tanzania, which was based on numbers
and prevalences of plague cases in villages across one district in the region
during 1986–2003. The authors first tested an initial suite of models to
identify ideal prevalence levels that best balanced omission and commission errors, and then presented detailed model evaluations over a local
landscape, where both presence and absence data were available. In the
end, they presented a risk map that was calibrated on the Usambara study
area but transferred to much of East Africa (figure 32). Although the details of their model calibration are satisfactory, the projections across East
Africa are not entirely appropriate, because the reader is not provided
with any information on the degree of novelty represented by the environ-

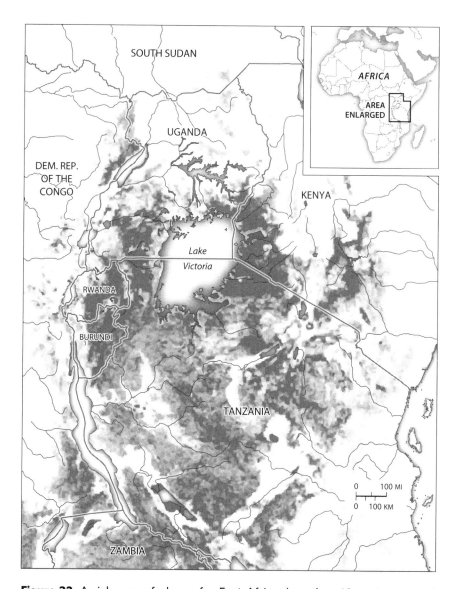

Figure 32. A risk map of plague for East Africa, based on 10 environmental variables and occurrences in 26 villages from Tanzania and the Democratic Republic of the Congo presenting moderate to high plague prevalences. Darker shades indicate areas with greater model agreement in predicting areas as being suitable to plague. From Neerinckx et al. 2009a.

ments across this broader region (Owens et al. 2013 discussed appropriate methodology). Another example of a black-box risk assessment was an analysis of Marburg virus distributions (Peterson et al. 2006a), notable for its careful management of differential uncertainty in georeferencing the few occurrences of the virus that were known at that time.

Component-Based Examples

In the simplest examples, a component-based approach may be merely a map of the distribution of vector species (or host species) that transmit a pathogen directly to humans. The transmission agent is the focus of such approaches, and other species that may be involved are either assumed to be ubiquitous (particularly if the association with those species is not strict), unknown, or overly complex to manage (e.g., Maher et al. 2010). These maps can be among the easiest and most feasible to develop, as they may depend on just a single (or small number of) species.

López-Cárdenas et al. (2005) presented fine-resolution mapping of the distributions of species of triatomine vectors of Chagas disease across the Mexican state of Guanajuato. The authors assembled a large-scale set of data characterizing occurrences of each species countrywide. When this study was conducted, only two triatomine specimens had previously been collected and reported from that state, so the authors used specimen records from surrounding states to calibrate models that could be transferred to Guanajuato. Field personnel from the state's Secretaría de Salud had conducted health promotion activities in 43 out of the 46 counties in the state and received donations of 2,522 triatomine specimens during 1998–2002, and the study's authors attempted to anticipate the spatial distributions of those many additional specimens from Guanajuato. The models, without exception, were highly predictive of the distributions of the species that were examined, and the study estimated that 3,755,380 people are at risk for vectorial transmission of Chagas disease in that state.

A study by Kulkarni et al. (2010) was somewhat more complete in representing a full, component-based transmission system. The authors developed niche model–derived maps of three important malaria vectors, based on remotely sensed land cover data sets and climatic data sets (figure 33). The niche models were highly accurate, but the critical test was whether they improved predictions of malaria prevalence in human populations. Integrating the presence of vector habitat within 1.5 km of human settlements with both community-based malaria prevalence measurements and data on elevation substantially improved predictions of the presence of *Plasmodium falciparum* in children. Work by Chamaillé et al. (2010) regarding canine leishmaniasis occurrences in southern France offered rich, component-based models and risk predictions. This study was particularly elegant in its careful use of hierarchical ascendant classification to group occurrences environmentally and, from them, to develop complementary models corresponding to two distinct environmental categories of occurrences.

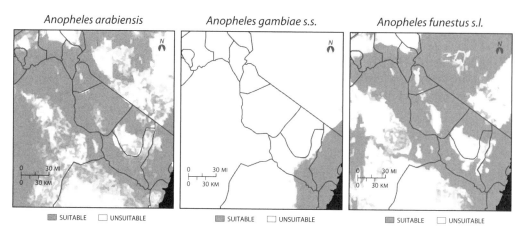

Figure 33. Areas identified as suitable (30 x 30 m resolution) across northeastern Tanzania for three dominant malaria vector species, *Anopheles arabiensis*, *A. gambiae sensu stricto*, and *A. funestus sensu lato.* From Kulkarni et al. 2010.

Improvements and Future Steps

Several additional steps still remain to be explored, in order to create the best and most predictive maps of disease transmission risk, and I emphasize two particularly crucial ones below. In both instances, good examples exist of what to do and what not to do, but best practices are not always possible or feasible or easy.

First, investigators should avoid the temptation to pile occurrence data and environmental data into a niche-modeling algorithm, press the button, and see what comes out. Rather, occurrence data must be assembled carefully and comprehensively, and biases, uncertainty, and temporal characteristics must be pondered. Once the input data are assembled, and the models calibrated appropriately (an important focus of this book), outputs become considerably more rigorous. Better and smarter tools and frameworks are needed, however, to transform raw model outputs into reliable risk maps. The most comprehensive goal is the development of entirely component-based models that will allow the broadest comparisons and considerations, and that will have been tested with fully independent data.

Second, ideal models should be based on remotely sensed data sets, rather than on climate data. In all but the broadest-extent applications, climate data lack sufficient detail to provide genuinely helpful information in public health applications. Under some circumstances, no alternatives are available, but satellite imagery is invariably richer in genuine information that is measured on real-world landscapes, rather than interpolated from frighteningly sparse weather station–based data. The development of better and smarter indices to transform raw imagery into relevant data layers that are the most informative to model development will also represent a significant step forward.

18

Identifying Species Involved in Transmission Cycles

Another key capacity that can be explored with ecological-biogeographic models of disease transmission risk is identifying unknown elements in the transmission cycles. While a few diseases have well-known cycles that have been characterized in enormous detail, many more have components that are unknown, merely suspected, or postulated without firm or sufficient evidence. In these latter cases, speculations about pathogens, vectors, and hosts are based on partial or anecdotal information. These disease mysteries can sometimes be clarified with ecological and geographic evidence, and such possibilities are the focus of this chapter.

Disease biology has necessarily been careful about conclusions regarding which species are responsible for disease. In 1900, Dr. James Carroll of the U.S. Army Yellow Fever Commission in Cuba wished to test a theory that yellow fever was transmitted by mosquitoes, so he let himself be bitten by mosquitoes (*Aedes aegypti*) that had previously fed on yellow fever patients. Carroll experienced a severe attack of yellow fever within a few days (although he recovered eventually). Carroll's technique for identifying elements of disease transmission cycles would certainly not be approved under modern biosafety protocols, but this example illustrates the lengths that sometimes have been necessary in tracking transmission vectors (Tabachnick 1991).

More commonly—and less recklessly—most of epidemiology has followed Koch's postulates and other, similar guidelines for conclusions about the elements in a disease transmission cycle (Plowright et al. 2008). Koch's postulates apply to pathogens: (1) the organism must be present when the disease is present, (2) the organism must not be associated with other diseases or present in normal tissues, and (3) the organism must be isolated from tissues in a pure culture and be capable of inducing disease under controlled experimental conditions. These criteria are rigid, but rigorous.

More recently, ecological evidence has been marshaled to illuminate disease transmission questions. Among the diversity of hosts involved in the transmission of Lyme disease in North America, the most susceptible is the broadly distributed mouse *Peromyscus leucopus*, but increased species diversity in these host communities appears to dilute their infection rates, because more ticks feed on inefficient disease reservoirs. This dilution effect was first documented via an appreciation of significant negative correlations between the species richness of small mammals and case rates of Lyme disease (Ostfeld and Keesing 2000). The evidentiary requirements here were not as rigorous as Koch's postulates, but the concept has been illustrative both for Lyme disease and a substantial variety of other diseases (Keesing et al. 2010).

Nonetheless, the details of transmission systems of many diseases remain cloudy. The extremely serious outbreak of SARS in the early 2000s is now known to have been caused by the SARS coronavirus. Initial conclusions were that the virus was hosted by civets (Guan et al. 2003), but they proved incorrect. The virus appears to be hosted instead by small bats (Li et al. 2005), but the details of its maintenance in those hosts, and the mechanisms by which it jumped to humans, remain entirely obscure. Aspects of many other diseases are also completely unknown. Nodding disease is a syndrome with a patchy distribution across parts of East Africa, with no information available about what causes it, the pathogen that might be responsible for it, other species involved in hosting and transmitting that pathogen, or whether the disease is even caused by a pathogen (Lacey 2003).

The realms of ecological and geographic information that are explored in this book can be informative about such situations, at least in some cases. Because Koch's first two postulates are correlational (Plowright et al. 2008), geographic and ecological parallels can be examined: whether a suspected organism's geographic range includes all sites where the disease is present, and whether that suspected organism's geographic range does not extend broadly into areas where the disease is not known (subject to the constraints of diagnosis, reporting, and socioeconomic factors, as well as of **B** in the BAM diagram). Hence hypotheses can be erected and tested regarding unknown elements of transmission systems by using ecological and geographic approaches.

Identifying Guilty Species

González et al. (2011) explored cutaneous leishmaniasis transmission across Mexico, where only *Lutzomyia olmeca olmeca* had been demonstrated to transmit *Leishmania mexicana* to humans. Although many other sandfly species are present in that country, their role in transmission of the disease was unknown, and no evidence was available to indicate that they were involved in transmitting *Leishmania* in Mexico. The authors compared the extent and limits of the geographic distribution of *Lu. o. olmeca* in relation to known leishmaniasis transmission areas in the country and found that the species' distribution was insufficient to account for all reported cases. In testing whether reported cases fell within the distribution of *Lu. o. olmeca*, they concluded that *Lu. cruciata* and *Lu. shannoni* represented additional potential vectors in Mexico. Such hypotheses, produced from geographic and ecological information, can then be explored and tested rigorously via other techniques, such as the isolation and characterization of parasites from the sandflies. A similar analysis identified additional likely vectors of malaria in humans in southeastern Brazil (Laporta et al. 2011).

Early analyses presented by Peterson et al. (2002b), focused on the *Triatoma protracta* species complex, their host associations, and the hosts' role in the transmission of Chagas disease. Following information presented in Ryckman (1962), Peterson and colleagues used the correspondence between distributional predictions for *Triatoma* parasites and those for their *Neotoma* hosts to test and document the close, one-to-one association of *Triatoma* species with *Neotoma* woodrats. For one *Triatoma* species (*T. barberi*), however, no host association information existed, but the authors' analysis noted a 98.4% distributional overlap with *N. mexicana*. This hypothesis regarding a possible host association with a Chagas disease vector has been the target of specific further testing by the same research group, but this has proven difficult, as both the bug and that species of woodrat are extremely uncommon. Although the range comparison approach of this early study has now been superseded by improved methodologies that explicitly weigh hypotheses of **M** (Warren et al. 2008; Petitpierre et al. 2012), this example is nonetheless still useful and relevant.

Understanding Transmission Systems

In other instances, the question is not so much the particular species that are involved in a transmission system, but the overall structure of the system and generalities about its function. For example, host and parasite (vector) associations in a disease transmission system could be either tight or loose, but these associations may not have been characterized in detail (Peterson et al. 2002b). Hence, generalities of this sort about disease transmission systems can be illuminated at times by ecological and geographic analyses.

Landscape-level studies of plague have identified local-scale risk factors that have considerable predictive abilities regarding where the potential for plague transmission exists (Parmenter et al. 1999; R. Eisen et al. 2007; Neerinckx et al. 2009a). On broader spatial scales, however, such as across continents, the ability to reconstruct plague transmission phenomena has been variable. Plague transmission is notoriously focal (Gage and Kosoy 2005), but various regional studies have been both able (Giles et al. 2010; Ben-Ari et al. 2012) and unable (Neerinckx et al. 2009b) to reconstruct and recover this focality, based on environment-distribution associations.

Given this relatively intractable system, a large-scale effort was made to build at least a partially component-based view of plague transmission, with substantial numbers of mammal plague case occurrences digitized to use in ecological and geographic comparisons with human case distributions. These disease components permitted novel comparisons. Maher et al. (2010) tested two competing hypotheses: the host niche hypothesis (in which the overall ecological and geographic potential is simply an aggregate of the ecological and geographic potentials of suites of host mammal species) versus the plague niche hypothesis (in which plague itself has a niche, and mammal species become infected with plague only where they overlap with those conditions and areas). Maher and colleagues used niche modeling to develop niche estimates for a suite of mammal species, in each case for the mammal in general as well as for the mammal when infected with plague. They then compared these niches using randomization tests. The evidence was quite strong toward the plague niche hypothesis, which probably indicates a role for flea species' distributions in shaping the plague niche, but data for North American flea distributions and occurrences are not available in digital formats.

Detecting Movement Vectors

West Nile virus arrived in the Americas in 1999 and spread rapidly across its new continents, reaching Argentina by 2006. Both experimental evidence (N. Komar et al. 2003) and evidence from mathematical models (Wonham et al. 2006) pointed to a rather broad spectrum of avian hosts. How the virus spread in the Americas was controversial, however, with different opinions claiming it was due to mosquito movements, bird migration, or human mediation (Rappole et al. 2000; Rappole and Hubálek 2003).

Peterson et al. (2003) used early patterns of the spread of the West Nile virus in the United States, which were well characterized by dead-bird monitoring programs (Eidson et al. 2001; N. Komar et al. 2002), as a basis for inference. In particular, Peterson and colleagues focused on an early disjunction that appeared in the eastern United States, which proved key in distinguishing between explanatory hypotheses. Initial detections (1999–2000) were concentrated in the northeastern United States, in concentric rings around the New York City area. By 2001, however, a new focus was established (in the southeastern United States, along the Georgia–Florida border) that was disjunct from the northeastern focus.

The authors used niche-modeling approaches to estimate distributions for relevant mosquito species and to derive range estimates for the breeding and wintering distributional areas for the relevant bird species. They then built component-based simulations that attempted to reflect how mosquitoes and locally resident birds might spread the virus, versus how migratory birds in tandem with local movements could disseminate it. Arguments at that time were that infected migratory birds would be too ill to fly long distances (Rappole and Hubálek 2003). Peterson et al. (2003), however, found clear evidence that the observed pattern of spread could only have been created with the participation of migratory birds, thus implicating these species in the geographic transmission cycle of West Nile virus.

Similar controversies arose a few years later with the Eastern Hemisphere–wide spread of a novel and highly pathogenic strain (H5N1) of avian influenza, which arose in Southeast Asia in the late 1990s and had extended west to Africa and Europe by around 2006. Numerous voices again stated that migratory birds would be too ill to carry this virus over long distances (Feare 2007), and many (particularly in the biodiversity conservation industry) argued that human-mediated movements of domestic birds (that is, bird trade) would prove responsible for the major features of the spread of H5N1 (BirdLife International 2006). A careful comparison of the ecological and geographic signatures of H5N1 influenza occurrence data across Europe, however, showed clear patterns of niche similarity between cases of this virus in wild and domestic birds, suggesting that transmission in the two sets of hosts is linked (Williams et al. 2011). These results concurred with other analyses, based on very different analytical frameworks, that had been developed by other research groups (Kilpatrick et al. 2006). Expected patterns of spread, based on migratory bird involvement, have now been characterized in detail, should H5N1 reach farther, such as into the

Americas (Peterson et al. 2007a; Peterson and Williams 2008; Peterson et al. 2009; Peterson, forthcoming).

Complete Unknowns

In some cases, diseases are so poorly known that any information relative to their transmission can be illuminating. An example is the filoviruses (Ebola and Marburg viruses), which were the subject of detailed analyses in this regard, although filovirus-host relationships were eventually discovered without reference to this body of research (Leroy et al. 2005). The following is a description of the lessons that were learned regarding this virus group; the overall body of evidence that was assembled may be informative and illustrative for efforts regarding other unknown disease systems.

A black-box mapping and niche-modeling effort provided initial, basic patterns (Peterson et al. 2004a), which involved a compilation of the geographic coordinates associated with occurrences of infections with these viruses, although the data were partial and incomplete, owing to the rare nature of transmission events in this system. These first analyses anticipated the geographic potential of the Marburg virus to reach as far south and west as Angola. This rather extreme hypothesis (at the time, the virus was not known west of Zimbabwe and only once south of Kenya, the Democratic Republic of the Congo, and Uganda) proved highly predictive, as a large outbreak occurred later that same year in northern Angola, precisely in the small portion of the country that the models had identified as suitable for transmission of the virus (Peterson et al. 2006a). The same research group then developed detailed analyses of the temporal dynamics of filovirus outbreaks (Lash et al. 2008), which provided greater detail and more rigor than had appeared in previous efforts along those lines (Pinzon et al. 2004).

The major effort in the filovirus work, however, focused on narrowing the list of mammal species that *could* participate in the transmission system as reservoir hosts of the virus. A first contribution along this line (Peterson et al. 2004b) compiled distributional data for a large number of African mammal species. The authors then made a series of explicit assumptions about what constituted a filovirus reservoir: (1) it is a mammal, (2) it supports persistent and largely asymptomatic filovirus infections, (3) its range subsumes that of associated filoviruses, (4) it has coevolved with the virus and therefore will show patterns of cophylogeny with it, (5) it has a small body size, and (6) it is not commensal with humans. This set of assumptions eliminated many species or clades of mammals as potential reservoirs for the viruses. The initial application of these assumptions taught several important lessons (Peterson et al. 2004b), and a second attempt—with much-improved distributional data (Peterson et al. 2007b)—resulted in a list of only 55 taxa that fit well with the assumption sets.

The final answer was on the list of potential reservoirs compiled by Peterson et al. (2007b), but it came from the en masse testing of bats for antibodies and viral genetic material (Leroy et al. 2005; Swanepoel et al. 2007; Towner et al. 2007, 2009). Even so, with clear indications that the bat *Rousettus aegyptiacus* is a long-term host for the filoviruses, many questions remain. Which bats host newly discovered filovirus lineages (Negredo et al. 2011)? What is the geographic distribution of the Ebola Reston virus and what is its possible long-term endemic status in the Philippines (Taniguchi et al. 2011)? Do two distinct, sympatric Marburg lineages (species?) exist in East Africa (Peterson and Holder 2012)? Host-virus associations among the seven to nine known filovirus species all present challenges to our current understanding.

Improvements and Future Steps

This chapter is perhaps the most fragmentary of the application chapters in this book. Few detailed examples have been developed to date as to how ecological and geographic methodologies can be applied to understanding the composition and functioning of disease transmission systems. The next major step forward that can and should be taken is that of developing, testing, exploring, and contemplating additional applications of these techniques. Only in this way will it be possible to assess the potential utility of these approaches.

Ecological and geographic applications will be richest when component-based summaries of disease transmission systems have been developed in greater detail, particularly when both component-based and black-box approaches can be assembled to estimate and understand the spatial filters involved in modifying actual transmission patterns and having them become reported patterns. Contrasts between these two views can illuminate not just the biases of diagnosis and reporting, but also the participation of additional species in a transmission system. The development of more-and-more detailed examples will be enormously helpful.

19

Responses to Environmental Change

The way species and disease transmission cycles are distributed at least partially in accord with environmental conditions suggests that distributions will be affected as environmental conditions change. The dimensions of those differences, however, have not been described definitively, and only effective model building and scenario exploration are likely to change that situation. Here, past and current efforts in approaching this question are reviewed, and potential steps forward are outlined and discussed.

Among the biggest unknowns in disease geography are the likely dimensions of how environmental change can affect geographic parameters. The impact of climate change on disease transmission became the object of speculation as early as the 1980s (Longstreth and Wiseman 1989), yet the first quantitative predictions regarding those effects did not appear until a decade later. Environmental change occurs in many dimensions and along several scales: climate change on decadal and global scales, land use change on shorter and finer scales, and seasonal dynamics on still shorter (and repeated) time scales.

If models for mapping disease transmission risk are to be genuinely useful and predictive, they must be able to anticipate the effects of such changes on spatial risk patterns. Tools for such inferences have been developed and explored extensively in biodiversity science. Approaches for inferring macrogeographic effects of climate change on species' distributions were developed in the late 1990s and early 2000s (Iverson et al. 1999;

Peterson et al. 2001, 2002a) and have now been explored in considerable detail (Dobrowski et al. 2010; Kearney et al. 2010). Methods for incorporating land use change in model outputs have also seen considerable exploration (Thuiller et al. 2004; Peterson et al. 2006c; Soberón and Peterson 2009).

Some disease systems have the advantage of process-based models, which made a first generation of environmental change studies possible (W. Martens et al. 1995; P. Martens et al. 1999). A second generation of disease-related projections has been based on empirical models (ecological niche models), with several published examples (Peterson 2009; González et al. 2010; Joyner et al. 2010). Third-generation thinking is now beginning to appear, in which dispersal considerations that go beyond the simple potential distributional projections of present-day solutions are also brought into the mix. Nonetheless, much work remains in this area, since robust scenarios of future disease transmission patterns have not been developed. Public health policy will benefit from a consistent, clear, and predictive set of forecasts for changes in disease transmission risk.

Early Mechanistic Models

The first generation of projections about the effects of environmental change on disease transmission areas was based primarily on mechanistic transmission models. Dengue fever is transmitted vectorially via one or two mosquito species in the genus *Aedes*, and Patz et al. (1998) linked a dengue vectorial capacity equation with climate data for present and modeled future conditions. The output from the models was a measure of epidemic potential. All of their future climate projections anticipated temperature-related increases in potential seasonal disease transmission in selected cities, as well as an increased possibility of a global epidemic; the largest areal change was in temperate regions. Hales et al. (2002) extended this early work with models based on dengue transmission as a function of vapor pressure, incorporating not only future climate changes, but also modeling variances in human population distributions. Their study indicated that the number of humans living in dengue-endemic regions is likely to increase three- to fourfold by 2085.

Ogden et al. (2005a) used a population model for the tick *Ixodes scapularis* (a vector of Lyme disease), with the model simulating population responses of the tick in the context of temperature variation through the year. This model was later applied to temperature data derived from two scenarios of climate change and used to anticipate alterations in the potential distribution of this tick species with future warming climates (Ogden et al. 2005b, 2008). The result was a clear picture of an advancing front of populations of these ticks into southern Canada in the coming decades (figure 34).

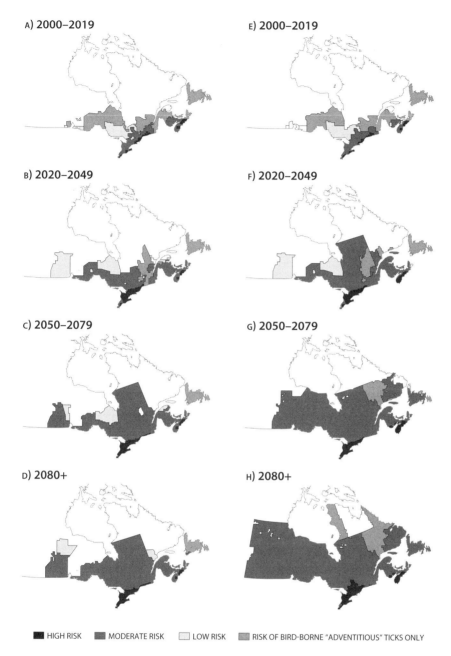

A) 2000–2019

E) 2000–2019

B) 2020–2049

F) 2020–2049

C) 2050–2079

G) 2050–2079

D) 2080+

H) 2080+

■ HIGH RISK   ■ MODERATE RISK   ☐ LOW RISK   ▨ RISK OF BIRD-BORNE "ADVENTITIOUS" TICKS ONLY

Figure 34. Maps summarizing the potential for occurrences of the Lyme disease vector *Ixodes scapularis* in terms of range expansion between the present and the 2080s (from the second version of the Coupled Global Climate Model [CGCM2], emissions scenario A2). Parts (a) through (d) summarize a "slow" scenario, while parts (e) through (h) summarize a "fast" scenario. From Ogden et al. 2008.

These early projections, based on mechanistic models, were instructive and outlined general patterns. They also faced some complex methodological challenges, however, such as that of scaling between the microclimates experienced by mosquitoes and ticks and the macroclimatic phenomena that make up coarse-resolution climate averages. In addition, while transmission models provide considerable detail, the degree to which they are broadly applicable is not clear. Transmission models may work well for a single vector species but probably do not succeed in capturing the complexity of multiple vector species that participate in some transmission cycles. They may also not capture geographic variations in the physiological functions of vector insects and pathogens. In this sense, empirical, data-driven models can provide a useful complement to the mechanistic perspective (Kearney et al. 2010).

Empirical Niche Model Projections of Climate Change

Second-generation evaluations of the effects of climate change on the geography of disease transmission risk relied on the broad, correlative power of ecological niche models. These studies began with an examination of leishmaniasis vectors in 2003 and have continued—with fast-increasing frequency—up to the present; more appear every month (e.g., Slater and Michael 2012). These approaches are more broadly available and applicable than mechanistic models, since the former do not depend on the existence of a transmission model and the associated assumptions that are required.

Leishmaniasis

Peterson and Shaw (2003) made early, niche-based climate change projections of the distributional potential of vector insects, focusing on *Lutzomyia* sandflies (vectors of cutaneous leishmaniasis across much of Brazil). They assembled occurrence data for each sandfly species in the region and developed niche model predictions using a genetic algorithm. The most striking was the suggestion that one species, *L. whitmani*, would potentially have a broad southward invasion into extreme southern Brazil (the states of Paraná, Santa Catarina, and Rio Grande do Sul), which is particularly significant in light of the changing epidemiology of this disease in South America (Oliveira et al. 2004). Those early projections are now being corroborated, with spotty records of *L. whitmani* having been documented across Paraná and apparently even south to Rio Grande do Sol (Massafera et al. 2005; S. Costa et al. 2007; Silva et al. 2008; Oliveira et al. 2011).

A more advanced, component-based model began to outline projections for the northern limit of leishmaniasis transmission in the Americas (González et al. 2010). Its focus was on two vector species (*Lutzomyia anthophora* and *L. diabolica*) and four likely rodent reservoir species (*Neotoma* spp.) for leishmaniasis in northern Mexico and the southern United States. Ecological niche models were calibrated for each of these six actors in the system. Model rule sets were then used to characterize present-day distributions of each species and were subsequently transferred to both conservative and liberal climate change projections (A2 and B2 emissions scenarios) for 2020, 2050, and 2080. The models projected a northward expansion of the potential ranges for both the vector and reservoir species, and the number of humans exposed to the disease is expected to double by 2080. Incorporating elevation as an independent variable in their models projected across changing climates, however, may cause inconsistencies in niche models (Peterson et al. 2011).

Malaria

A long sequence of studies and debates has addressed the role that climate change may or may not play in the transmission of malaria. An early projection was based on a transmission model for *Plasmodium falciparum* malaria in Africa and was applied to different climate scenarios, which anticipated a 5%–7% increase in the potential distribution of this disease, with only minimal macrogeographic implications (Tanser et al. 2003).

The focus on malaria and climate change then shifted to higher-elevation areas of East Africa. The International Panel on Climate Change anticipated a distributional expansion and increased incidences of malaria in the region. Hay et al. (2002) analyzed long-term meteorological trends at four high-elevation sites where malaria increases had been reported in the preceding two decades and concluded that climate parameters and suitability for *Plasmodium falciparum* malaria transmission had not changed appreciably. Pascual et al. (2006), however, using the same data sets, found evidence for significant warming trends at all sites. Tying this refined information to a dynamic vector population model, they found that mosquito population dynamics would amplify much more dramatically than the climate shifts, suggesting that climate parameters indeed do play a role in increases in the incidence of malaria in the region.

Peterson (2011) took a broader-scope view and analyzed continent-wide distributions of two key malaria vectors (*Anopheles gambiae* and *A. arabiensis*) with regard to climate parameters. Potential distributional shifts in relation to climate-model-based projections of future climate conditions were assessed, and the study then focused on changing numbers of humans living in regions that would be climatically suitable for these vector species over the coming 50 years. For both vector species, under all scenarios, the models anticipated an expansion of the malaria transmission area into southern and eastern Africa, but a retraction along the southern rim of the Sahara Desert, particularly in West Africa. As a consequence of uneven human distributions on the continent, the models predicted a 11.3%–30.2% reduction in the number of people in Africa living in areas climatically suitable for these vector species under future conditions.

Dengue

In recent decades, the mosquito species *Aedes albopictus* has spread globally, providing another potential major vector of diseases such as dengue fever. Fischer et al. (2011b) explored four niche-modeling approaches to evaluate its global distributional potential, selecting climate variables based on (1) expert knowledge versus (2) statistical criteria, and calibrating models based on (3) the native range of the species versus (4) across its global range. Native-range models showed good performance but poor transferability, since they failed to anticipate other distributional areas that were already colonized. Models calibrated across the global distribution of the species, however, were more successful in their identification of the species' European range. Transferring these models onto future climatic conditions in Europe (using two emission scenarios implemented in a regional climate model for the years 2011–2040, 2041–2070, and 2071–2100), the species was projected to see expanding distributional possibilities in western and central Europe early in these periods, and later in eastern Europe.

More detailed analyses of dengue have focused on vector responses to variations in the environment. In subtropical regions, seasonal dynamics present significant temporal environmental changes for short-lived mosquito species. Peterson et al. (2005) used niche-modeling approaches to relate monthly occurrence data for *Aedes aegypti* to remotely sensed data summarizing local, month-specific environmental conditions. They found rather complex spatial dynamics, but ecological niche-modeling approaches could be used with considerable confidence to predict and anticipate when and for how long the mosquitoes would appear. The authors then evaluated how the distributional potential of the mosquitoes correlated with cases of dengue in humans. The model anticipated—with good statistical significance—human case rates appearing 18 days later (the time for a mosquito to bite an infected human, incubate the virus, and infect another human, and for the symptoms to begin), suggesting that the dynamics of vector distribution can translate into a human infection risk.

Ovine Cutaneous Myiasis

This veterinary disease is caused by the larvae of a parasitic blowfly (most commonly *Lucilia sericata* in northern Europe), with severe economic consequences. Rose and Wall (2011) set out to characterize the environmental determinants of monthly occurrences of this disease and the variation in occurrences over the course of the year. They then used the model to assess geographic shifts in the risk of infection that might be expected as a function of climate change across Great Britain (which would affect seasonal temperature variations). The authors identified a range of elevated temperatures that would be key in increasing infection risk and extending the blowfly season, although transmission probably would not be continuous year round, and parts of central and southern England might become too hot and dry for this fly species to persist in midsummer. The authors went on to suggest changes in the timing and frequency of parasite treatments and husbandry practices, such as shearing, to alleviate present and future risks for this disease in sheep.

Anthrax

Joyner et al. (2010) used a large data set of occurrences to develop ecological niche models and assess changing climate footprints across Kazakhstan in relation to anthrax, caused by the bacterium *Bacillus anthracis*. They explored climate change implications for anthrax transmission under two scenarios of future climate conditions. Their results anticipated some reduction in the potential transmission region by 2050, and only small areas of distributional expansion. A concern regarding this study—unavoidable in light of the restricted data availability—is that the niche of the bacterium may have been only partially characterized because of the limited geographic scope of **S**, compared with the species' worldwide **M**.

Mechanistic versus Empirical Models

Mechanistic models were used initially in mapping changes in disease transmission risk, in tandem with some of the earlier, general circulation model (GCM) climate projections. Later approaches have focused primarily on data-driven, empirical models (niche models) and have been applicable to a broader diversity of diseases. In light of their differences, perhaps some contemplation of the relative merits and drawbacks of the two approaches is in order.

Mechanistic models provide amazing detail on the environmental influences that permit or constrain the transmission of diseases, but few have been developed (presently only malaria, dengue, and just a few others), which greatly limits the number of diseases to which they can be applied.

Moreover, these models do not necessarily capture all of the complexity of a disease transmission system (e.g., alternative vectors) or its details, such as possible physiological variation across the distributional areas of vector species. The final, and perhaps most challenging, aspect is that transmission models are frequently (and correctly) cast on microclimatic resolutions. Translating microclimates into macroclimatic conditions that can be mapped often will involve massive assumption sets and miss microclimatic refuges.

In the biodiversity world, recent comparisons have generally found good agreement between mechanistic and correlative (niche) models (Kearney et al. 2010). The niche-modeling approach is empirical and data driven, and thus is able to take more of the intricacy of complex transmission systems into account. Nonetheless, niche models must often be constrained to small **M ∩ S** areas that are accessible to the species and have been sampled adequately, such that in many situations researchers may not be able to fit complete niche models. Niche models for anticipating responses to environmental change may be particularly liable to challenges when they are black-box models of complex transmission systems, since such models assume that interspecific relationships will remain constant when they may, in fact, change.

Improvements and Future Steps

The most notable feature of this chapter should be the relatively small number of disease systems—the case studies listed above—to which mechanistic and empirical techniques have been applied. Different vector-borne flaviviruses, as well as Chagas disease, tick-borne diseases, and others, would lend themselves well to these approaches, but they have not yet been assessed. This gap is significant in light of public health needs to adapt strategically to the changing geography of disease transmission, such as that produced by current climate changes.

The applications that have been developed—and those to be developed in the future—would benefit from a deeper and more detailed consideration of several factors. For climate change questions, surveys across multiple climate models would remove the model-to-model variation that is inherent in all of the examples developed to date. Similarly, taking the full complement of the Intergovernmental Panel on Climate Change's emissions scenarios (IPCC 2007) into account would enrich the insights to be garnered from studies of climate change. Perhaps most importantly, such studies should carefully assess the degree of uncertainty in model results: uncertainties derived from the climate models, from occurrence data sampling, and from model transfers (using MOP-like approaches).

Finally, empirical models could be used in developing and testing hypotheses about important factors in changing disease transmission patterns. Nakazawa et al. (2007) tested the probable role of the effects of climate shifts in molding the geography of tularemia and plague transmissions across North America, finding that—in both cases—observed range shifts for these diseases in recent periods are consistent with how climates are changing over the past several decades and into the future. More tests of this sort can be instrumental in illuminating the environmental factors that do—or do not—shape disease transmission across broad geographic extents.

20

Conclusions

Mapping disease transmission risk—the subject of this book—is an objective that has been pursued for more than a century. The methods used for this purpose, however, have not evolved as quickly as might be hoped. Until relatively recently, disease transmission geography was treated as a purely spatial phenomenon, and the ecological and biogeographic contexts of disease transmission were ignored or forgotten. Currently, environmental information has been incorporated increasingly in mapping efforts, but without an explicit conceptual framework to guide the analyses. This book has sought to fill that gap.

The challenge of mapping disease distributions (and, by extension, disease transmission risk) has been a focus in public health for centuries (T. Koch 2005). From the earliest disease maps, which simply plotted occurrences, to recent efforts using complex inferential approaches to incorporate the effects of human population, spatial location, and other confounding factors (Lawson 2008), these efforts have treated disease geography as a spatial phenomenon. Most mapping efforts to date have incorporated mainly spatial information, and they generally have not considered the environmental variations manifested across the landscape in question.

Real-world landscapes, however, present a dramatic environmental diversity that undoubtedly constrains and shapes the geography of disease transmission. On the broadest and coarsest scales, climates vary from tropical to arctic, with complex, scale-dependent effects that challenge even basic efforts to map climatic variation (much less disease transmission risk). On more refined scales, land use, vegetation types, habitat fragmentation, soil composition, topographic geometry, and many other factors present finer-grained variations across landscapes that make simple, space-only approaches less and less useful. Finally, because most of the available data sets that characterize disease distributions are highly uneven across space, significant biases emerge. Space-only approaches will readily confuse patchiness in sampling with concentrations of disease transmission.

New initiatives have begun to take on some of the significant challenges in disease mapping. Diseases that are the foci of major efforts include malaria (Moyes et al. 2013) and dengue (Bhatt et al. 2013), and researchers are now eyeing the broader challenge of simply displaying everything, such as the recent evaluation of the "mappability" of "all infectious diseases of clinical significance to humans" (Hay et al. 2013). This book, I hope, has offered a better methodological structure in which these mapping challenges can and should be cast. Without the guiding conceptual framework of ecology and biogeography, such maps will frequently misrepresent true transmission risks, as can be appreciated from the example of mapping Lassa fever risk (Fichet-Calvet and Rogers 2009) that was presented in chapter 4.

The field as it currently stands is moving fast, but it has a number of failings that will limit its long-term progress. These infrastructural challenges include a culture centered on data-hoarding rather than data-sharing, non-standardized data formats when data *are* available, and samples that are poorly documented in terms of metadata and voucher specimens. In short, the information sets that underlie all of the analyses one might develop are poorly managed, without adequate consideration given to archival storage, efficient sharing and distribution methods, and maximized utility (by recycling and reusing relevant data for multiple analyses). These challenges have been addressed in recent decades in the biodiversity world, but the public health arena lags behind. Such problems are neither exciting nor "sexy," but in the long term they will determine the forward progress—or lack thereof—in this latter field.

Apart from data infrastructures, the methodologies that have been explored and explained in this book are far from mature and complete. In my view, major challenges will center on (1) explicitly, quantitatively, repeatedly, and effectively estimating **M** areas for the species in question; (2) seeking out appropriate and optimal response surfaces that should be estimated by modeling algorithms, since complex responses are not necessarily the best ones; and (3) incorporating and fully integrating dispersal considerations within modeling and mapping frameworks (an excellent example of possibilities in this realm can be found in Fischer et al. 2011a). None of these challenges will be resolved easily or quickly, but each will create qualitative improvements in the mapping framework presented here.

One prospect specifically related to the topic of this volume is an exploration of the potentially massive insights that most likely will emerge from the integration of component-based and black-box approaches to disease mapping: one documenting the components of the systems that interact to transmit diseases, and the other documenting the end products of those systems, which are disease cases among humans and their associated animals. These two means of examining disease distributions are already available, and this book has discussed the numerous data challenges accompanying their integration. Getting component-based and black-box data assembled for the same disease system over the same region is not easy, and frequently it will not be feasible. Yet, as the numerous filters and biases that translate between the two views become explicit, multiple new insights will emerge.

A final move forward will be to shift the entire culture of this realm of science, not just its methodological details. Scientific disciplines construct frameworks of how to do business that are resistant to change. Levine et al. (2004b) applied niche-modeling approaches to *Anopheles* mosquitoes across Africa, yet Coetzee (2004), a leader in this field, stated:

> There is no doubt that geographic information systems and distribution modeling are useful components of an efficient malaria control program where models are based on verified distributional data. They are best used at the regional level (www.malaria.org) for practical program management rather than on a continent-wide scale where resolution is coarse and applicability limited.

Evolution and improvements in this culture do not come easily, yet such shifts do happen. One example is the change from space-only to space-and-environment modeling of disease transmission geography. Figure 35 summarizes trends in scientific journal articles that link the word "disease" with the terms "ecological niche model," "species distribution model," "Maxent," or "GARP," based on a Web of Science search. (Space-and-environment models are more numerous than the quantities in this figure, but it gets the basic message across.) The temporal trend is striking: these methodologies are being adopted with increasing frequency in the field, and disease workers are exploring their possible utility.

Still, many steps remain. In the world of disease mapping, just using the tools does not necessarily mean that the correct underlying frameworks are

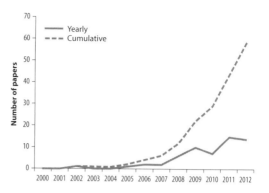

Figure 35. A summary of trends since 2000 in the publication of papers referring to "disease" and any of the following phrases: "ecological niche model," "species distribution model," "Maxent," or "GARP." The two curves present yearly total numbers and cumulative totals.

being employed. Merely assembling occurrence data for relevant species and environmental data for the landscape, and then "pressing the button," does not produce an effective and predictive model. Rather, one's initial thoughts in approaching an issue, and the resultant analyses, can and should be guided by the appropriate frameworks. Ecology can help translate the physiology and natural history of a species into coarse-grained environmental tolerances. Similarly, biogeography can offer a way to show how species' distributions extend across present-day and historical landscapes in the real world. These conceptual guides can make a substantial difference between supposed "models" that simply respond to input data (remember the garbage-in-garbage-out principle) versus models that have significant predictive power.

In the end, the question is not who is right, or which method is best. The goal is to evolve a methodology that allows workers in the field to develop a predictive and informative understanding of where disease transmission events are likely to occur. A geographic view of disease transmission risk can guide public health efforts in intervention and mitigation; the results can be improved human and animal health, and greater cost-efficiency in health-related spending. Refining and expanding the linkages between science, policy, and public well-being can be of immense benefit to all concerned.

Literature Cited

Adhikari, D., A. Chettri, and S. K. Barik. 2009. Modelling the ecology and distribution of highly pathogenic avian influenza (H5N1) in the Indian subcontinent. Current Science 97:73–78.

Aidoo, M., D. J. Terlouw, M. S. Kolczak, P. D. McElroy, F. O. ter Kuile, S. Kariuki, B. L. Nahlen, A. A. Lal, and V. Udhayakumar. 2002. Protective effects of the sickle cell gene against malaria morbidity and mortality. Lancet 359:1311–1312.

Allouche, O., A. Tsoar, and R. Kadmon. 2006. Assessing the accuracy of species distribution models: Prevalence, kappa and the true skill statistic (TSS). Journal of Applied Ecology 43:1223–1232.

Anderson, R. P. 2003. Real vs. artefactual absences in species distributions: Tests for *Oryzomys albigularis* (Rodentia: Muridae) in Venezuela. Journal of Biogeography 30:591–605.

Anderson, R. P., D. Lew, and A. T. Peterson. 2003. Evaluating predictive models of species' distributions: Criteria for selecting optimal models. Ecological Modelling 162:211–232.

Andreasen, V., J. Lin, and S. A. Levin. 1997. The dynamics of cocirculating influenza strains conferring partial cross-immunity. Journal of Mathematical Biology 35:825–842.

Antolin, M. F., P. Gober, B. Luce, D. E. Biggins, W. E. V. Pelt, D. B. Seery, M. Lockhart, and M. Ball. 2002. The influence of sylvatic plague on North American wildlife at the landscape level, with special emphasis on black-footed ferret and prairie dog conservation. Pages 104–127 *in* Transactions of the Sixty-Seventh North American Wildlife and Natural Resources Conference (J. Rahm, ed.). North American Wildlife and Natural Resources Conference, Washington, DC.

Arboleda, S., N. Jaramillo-O., and A. T. Peterson. 2009. Mapping environmental dimensions of dengue fever transmission risk in the Aburrá Valley, Colombia. International Journal of Environmental Research and Public Health 6:3040–3055.

Arias, A. R. d., F. Abad-Franch, N. Acosta, E. López, N. González, E. Zerba, G. Tarelli, and H. Masuh. 2012. Post-control surveillance of *Triatoma infestans* and *Triatoma sordida* with chemically baited sticky traps. PLoS Neglected Tropical Diseases 6:e1822.

Ariño, A. H. 2010. Approaches to estimating the universe of natural history collections data. Biodiversity Informatics 7:81–92.

Ashley, S. T., and V. Meentemeyer. 2004. Climatic analysis of Lyme disease in the United States. Climate Research 27:177–187.

Ashraf, M. 2004. Some important physiological selection criteria for salt tolerance in plants. Flora 199:361–376.

Avise, J. C. 2000. Phylogeography: The History and Formation of Species, 3rd ed. Harvard University Press, Cambridge, MA.

Badaró, R., T. C. Jones, R. Lorenço, B. J. Cerf, D. Sampaio, E. M. Carvalho, H. Rocha, R. Teixeira, and W. D. Johnson. 1986. A prospective study of visceral leishmaniasis in an endemic area of Brazil. Journal of Infectious Diseases 154:639–649.

Bar-Zeev, M. 1960. The reaction of mosquitoes to moisture and high humidity. Entomologia Experimentalis et Applicata 3:198–211.

Barrett, A. D. T., and S. Higgs. 2007. Yellow fever: A disease that has yet to be conquered. Annual Review of Entomology 52:209–229.

Barve, N., V. Barve, A. Jiménez-Valverde, A. Lira-Noriega, S. P. Maher, A. T. Peterson, J. Soberón, and F. Villalobos. 2011. The crucial role of the accessible area in ecological niche modeling and species distribution modeling. Ecological Modelling 222:1810–1819.

Bausch, D. G., M. Borchert, T. Grein, C. Roth, R. Swanepoel, M. L. Libande, A. Talarmin, E. Bertherat, J. J. Muyembe-Tamfum, B. Tugume, R. Colebunders, K. M. Konde, P. Pirard, L. L. Olinda, G. R. Rodier, P. Campbell, O. Tomori, T. G. Ksiazek, and P. E. Rollin. 2003. Risk factors for Marburg hemorrhagic fever, Democratic Republic of the Congo. Emerging Infectious Diseases 9:1531–1537.

Bayoh, M. N., and S. W. Lindsay. 2004. Temperature-related duration of aquatic stages of the Afrotropical malaria vector mosquito *Anopheles gambiae* in the laboratory. Medical and Veterinary Entomology 18:174–179.

Beale, C. M., J. J. Lennon, and A. Gimona. 2008. Opening the climate envelope reveals no macroscale associations with climate in European birds. Proceedings of the National Academy of Sciences USA 105:14908–14912.

Beard, C. B., G. Pye, F. J. Steurer, Y. Salinas, R. Campman, A. T. Peterson, J. M. Ramsey, R. A. Wirtz, and L. E. Robinson. 2002. Chagas disease in a domestic transmission cycle in southern Texas, USA. Emerging Infectious Diseases 9:103–105.

Ben-Ari, T., S. Neerinckx, L. Agier, B. Cazelles, L. Xu, Z. Zhang, X. Fang, S. Wang, Q. Liu, and N. C. Stenseth. 2012. Identification of Chinese plague foci from long-term epidemiological data. Proceedings of the National Academy of Sciences USA 109:8196–8201.

Benedict, M. Q., R. S. Levine, W. A. Hawley, and L. P. Lounibos. 2007. Spread of the tiger: Global risk of invasion by the mosquito *Aedes albopictus*. Vector-Borne and Zoonotic Diseases 7:76–85.

Bhatt, S., P. W. Gething, O. J. Brady, J. P. Messina, A. W. Farlow, C. L. Moyes,

J. M. Drake, J. S. Brownstein, A. G. Hoen, and O. Sankoh. 2013. The global distribution and burden of dengue. Nature 496:504–507.

Bi, Z., P. Formenty, and C. E. Roth. 2008. Hantavirus infection: A review and global update. Journal of Infection in Developing Countries 2:3–23.

Birch, L. C. 1953. Experimental background to the study of the distribution and abundance of insects: I. The influence of temperature, moisture and food on the innate capacity for increase of three grain beetles. Ecology 34:698–711.

BirdLife International. 2006. BirdLife statement on avian influenza. BirdLife International, Cambridge. www.birdlife.org/action/science/species/avian_flu/ [no longer available].

Blum, S., D. Vieglais, and P. J. Schwartz. 2001. DiGIR—Distributed Generic Information Retrieval, version 1.5. SourceForge, Inc.

Bodbyl-Roels, S., A. T. Peterson, and X. Xiao. 2011. Comparative analysis of remotely sensed data products via ecological niche modeling of avian influenza case occurrences in Middle Eastern poultry. International Journal of Health Geographics 10:21.

Bolaños, J., G. O. Edmeades, and L. Martinez. 1993. Eight cycles of selection for drought tolerance in lowland tropical maize: III. Responses in drought-adaptive physiological and morphological traits. Field Crops Research 31:269–286.

Brault, A. C., S. A. Langevin, R. A. Bowen, N. A. Panella, B. J. Biggerstaff, B. R. Miller, and N. Komar. 2004. Differential virulence of West Nile strains for American Crows. Emerging Infectious Diseases 10:2161–2168.

Breiman, L. 2001. Random forests. http://oz.berkeley.edu/users/breiman/random forest2001.pdf. Unpublished technical report, University of California, Berkeley.

Broennimann, O., U. A. Treier, H. Müller-Schärer, W. Thuiller, A. T. Peterson, and A. Guisan. 2007. Evidence of climatic niche shift during biological invasion. Ecology Letters 10:701–709.

Brooker, S., S. I. Hay, and D. A. P. Bundy. 2002. Tools from ecology: Useful for evaluating infection risk models? Trends in Parasitology 18:70–74.

Brown, H. E., K. F. Yates, G. Dietrich, K. MacMillan, C. B. Graham, S. M. Reese, W. S. Helterbrand, W. L. Nicholson, K. Blount, P. S. Mead, S. L. Patrick, and R. J. Eisen. 2011. An acarologic survey and *Amblyomma americanum* distribution map with implications for tularemia risk in Missouri. American Journal of Tropical Medicine and Hygiene 84:411–419.

Brownstein, J. S., T. R. Holford, and D. Fish. 2003. A climate-based model predicts the spatial distribution of the Lyme disease vector *Ixodes scapularis* in the United States. Environmental Health Perspectives 111:1152–1157.

Camargo-Neves, V. L., C. Gomes Ade, and J. L. Antunes. 2002. Correlação da presença de espécies de flebotomíneos (Diptera: Psychodidae) com registros de casos da leishmaniose tegumentar americana no Estado de São Paulo, Brasil. Revista da Sociedade Brasileira de Medicina Tropical 35:299–306.

Carpenter, G., A. N. Gillison, and J. Winter. 1993. DOMAIN: A flexible modeling procedure for mapping potential distributions of animals and plants. Biodiversity and Conservation 2:667–680.

Castro-Arellano, I., G. Suzán, R. F. León, R. M. Jiménez, and T. E. Lacher Jr. 2009. Survey for antibody to hantaviruses in Tamaulipas, Mexico. Journal of Wildlife Diseases 45:207–212.

Chamaillé, L., A. Tran, A. Meunier, G. Bourdoiseau, P. Ready, and J.-P. Dedet. 2010. Environmental risk mapping of canine leishmaniasis in France. Parasites and Vectors 3:31.

Chapman, A. D. 2005a. Principles and Methods of Data Cleaning, version 1.0. Global Biodiversity Information Facility, Copenhagen.

Chapman, A. D. 2005b. Principles of Data Quality, version 1. Global Biodiversity Information Facility, Copenhagen.

Chapman, A. D., and J. Wieczorek, eds. 2006. Guide to best practices for georeferencing. Global Biodiversity Information Facility, Copenhagen.

Chase, J. M., and M. A. Leibold. 2003. Ecological Niches: Linking Classical and Contemporary Approaches. University of Chicago Press, Chicago.

Chu, Y. K., R. D. Owen, C. Sánchez-Hernández, M. L. Romero-Almaraz, and C. B. Jonsson. 2008. Genetic characterization and phylogeny of a hantavirus from western Mexico. Virus Research 131:180–188.

Coetzee, M. 2004. Distribution of the African malaria vectors of the *Anopheles gambiae* complex. American Journal of Tropical Medicine and Hygiene 70:103–104.

Coetzee, M., M. Craig, and D. le Sueur. 2000. Distribution of African malaria mosquitoes belonging to the *Anopheles gambiae* complex. Parasitology Today 16:74–77.

Colwell, R. K., and J. A. Coddington. 1994. Estimating terrestrial biodiversity through extrapolation. Philosophical Transactions of the Royal Society of London B 335:101–118.

Costa, J. 1999. The synanthropic process of Chagas disease vectors in Brazil, with special attention to *Triatoma brasiliensis* Neiva, 1911 (Hemiptera, Reduviidae, Triatominae) population, genetical, ecological, and epidemiological aspects. Memórias do Instituto Oswaldo Cruz 94:239–241.

Costa, J., A. T. Peterson, and C. B. Beard. 2002. Ecological niche modeling and differentiation of populations of *Triatoma brasiliensis* Neiva, 1911, the most important Chagas disease vector in northeastern Brazil (Hemiptera, Reduviidae, Triatominae). American Journal of Tropical Medicine and Hygiene 67:516–520.

Costa, S. M. de, M. Cechinel, V. Bandeira, J. C. Zannuncio, R. Lainson, and E. F. Rangel. 2007. *Lutzomyia* (*Nyssomyia*) *whitmani s.l.* (Antunes & Coutinho, 1939) (Diptera: Psychodidae: Phlebotominae): Geographical distribution and the epidemiology of American cutaneous leishmaniasis in Brazil. Memórias do Instituto Oswaldo Cruz 102:149–153.

Cromley, E. K., and S. L. McLafferty. 2002. GIS and Public Health. Guilford Press, New York.

Cully, J. F., A. M. Barnes, T. J. Quan, and G. Maupin. 1997. Dynamics of plague in a Gunnison's prairie dog colony complex from New Mexico. Journal of Wildlife Diseases 33:706–719.

Damon, I. K., C. E. Roth, and V. Chowdhary. 2006. Discovery of monkeypox in Sudan. New England Journal of Medicine 355:962–963.

Daszak, P., A. A. Cunningham, and A. D. Hyatt. 2000. Emerging infectious diseases of wildlife: Threats to biodiversity and human health. Science 287:443–449.

Day, J. F. 2001. Predicting St. Louis encephalitis virus epidemics: Lessons from recent, and not so recent, outbreaks. Annual Review of Entomology 46:111–138.

Dias, J. C. P., A. C. Silveira, and C. J. Schofield. 2002. The impact of Chagas disease control in Latin America: A review. Memórias do Instituto Oswaldo Cruz 97:603–612.

Dobrowski, S. Z., J. H. Thorne, J. A. Greenberg, H. D. Safford, A. R. Mynsberge, S. M. Crimmins, and A. K. Swanson. 2010. Modeling plant ranges over 75 years of climate change in California, USA: Temporal transferability and species traits. Ecological Monographs 81:241–257.

Donalisio, M. R., and A. T. Peterson. 2011. Environmental factors affecting transmission risk for hantaviruses in forested portions of southern Brazil. Acta Tropica 119:125–130.

Dupuis, A. P., II, P. P. Marra, and L. D. Kramer. 2003. Serologic evidence of West Nile virus transmission, Jamaica, West Indies. Emerging Infectious Diseases 9:860–863.

Dupuis, A. P., II, P. P. Marra, R. Reitsma, M. J. Jones, K. L. Louie, and L. D. Kramer. 2005. Serologic evidence for West Nile virus transmission in Puerto Rico and Cuba. American Journal of Tropical Medicine and Hygiene 73:474–476.

Eidson, M., L. D. Kramer, W. Stone, Y. Hagiwara, K. Schmitt, and New York State West Nile Virus Avian Surveillance Team. 2001. Dead bird surveillance as an early warning system for West Nile virus. Emerging Infectious Diseases 7:631–635.

Eisen, L., and R. J. Eisen. 2007. Need for improved methods to collect and present spatial epidemiologic data for vector-borne diseases. Emerging Infectious Diseases 13:1816–1820.

Eisen, L., and R. J. Eisen. 2008. In response. Emerging Infectious Diseases 14: 1336–1337.

Eisen, R. J., R. E. Enscore, B. J. Biggerstaff, P. J. Reynolds, P. Ettestad, T. Brown, J. Pape, D. Tanda, C. E. Levy, D. M. Engelthaler, J. Cheek, R. Bueno, J. Targhetta, J. A. Montenieri, and K. L. Gage. 2007. Human plague in the southwestern United States, 1957–2004: Spatial models of elevated risk of human exposure to *Yersinia pestis*. Journal of Medical Entomology 44:530–537.

Eisen, R. J., R. S. Lane, C. L. Fritz, and L. Eisen. 2006. Spatial patterns of Lyme disease risk in California based on disease incidence data and modeling of vector-tick exposure. American Journal of Tropical Medicine and Hygiene 75:669–676.

Elith, J., and C. H. Graham. 2009. Do they? How do they? WHY do they differ? On finding reasons for differing performances of species distribution models. Ecography 32:66–77.

Elith, J., C. H. Graham, R. P. Anderson, M. Dudík, S. Ferrier, A. Guisan, R. J. Hijmans, F. Huettmann, J. R. Leathwick, A. Lehmann, J. Li, L. G. Lohmann, B. A. Loisell, G. Manion, C. Moritz, M. Nakamura, Y. Nakazawa, J. Overton, A. T. Peterson, S. J. Phillips, K. Richardson, R. Scachetti-Pereira, E. Schapire, J. Soberón, S. Williams, M. S. Wisz, and N. E. Zimmerman. 2006. Novel methods improve prediction of species' distributions from occurrence data. Ecography 29:129–151.

Elith, J., M. Kearney, and S. Phillips. 2010. The art of modelling range-shifting species. Methods in Ecology and Evolution 1:330–342.

Elith, J., S. J. Phillips, T. Hastie, M. Dudík, Y. E. Chee, and C. J. Yates. 2011. A statistical explanation of MaxEnt for ecologists. Diversity and Distributions 17:43–57.

Ellis, C. K., D. S. Carroll, R. R. Lash, A. T. Peterson, I. K. Damon, J. Malekani, and P. Formenty. 2012. Ecology and geography of human monkeypox case occurrences across Africa. Journal of Wildlife Diseases 48:335–347.

Elton, C. S. 1927. Animal Ecology. Sidgwich and Jackson, London.

Enria, D., P. Padula, E. L. Segura, N. Pini, A. Edelstein, C. Riva Posse, and M. C. Weissenbacher. 1996. Hantavirus pulmonary syndrome in Argentina: Possibility of person to person transmission. Medicina 56:709–711.

Estrada-Franco, J. G., R. Navarro-López, D. W. C. Beasley, L. Coffey, A.-S. Carrara, A. Travassos da Rosa, T. Clements, E. Wang, G. V. Ludwig, A. Campomanes-Cortes, P. Paz-Ramírez, R. B. Tesh, A. D. T. Barrett, and S. C. Weaver. 2003. West Nile virus in Mexico: Evidence of widespread circulation since July 2002. Emerging Infectious Diseases 9:1604–1607.

Estrada-Peña, A., Z. Zatansever, A. Gargili, M. Aktas, R. Uzun, O. Ergonul, and F. Jongejan. 2007. Modeling the spatial distribution of Crimean-Congo hemorrhagic fever outbreaks in Turkey. Vector-Borne and Zoonotic Diseases 7:667–678.

Evans, M. E. K., S. A. Smith, R. S. Flynn, and M. J. Donoghue. 2009. Climate, niche evolution, and diversification of the "bird-cage" evening primroses (*Oenothera*, sections *Anogra* and *Kleinia*). American Naturalist 173:225–240.

Feare, C. J. 2007. The role of wild birds in the spread of HPAI H5N1. Avian Diseases 51:440–447.

Ferrier, S., G. Watson, J. Pearce, and M. Drielsma. 2002. Extended statistical approaches to modelling spatial pattern in biodiversity in northeast New South Wales: I. Species-level modelling. Biodiversity and Conservation 11:2275–2307.

Fichet-Calvet, E., and D. J. Rogers. 2009. Risk maps of Lassa fever in West Africa. PLoS Neglected Tropical Diseases 3:e388.

Field, H. E., J. S. Mackenzie, and P. Daszak. 2007. Henipaviruses: Emerging paramyxoviruses associated with fruit bats. Current Topics in Microbiology and Immunology 315:133–159.

Fielding, A. H., and J. F. Bell. 1997. A review of methods for the assessment of prediction errors in conservation presence/absence models. Environmental Conservation 24:38–49.

Fischer, D., S. M. Thomas, and C. Beierkuhnlein. 2011a. Modelling climatic suitability and dispersal for disease vectors: The example of a phlebotomine sandfly in Europe. Procedia Environmental Sciences 7:164–169.

Fischer, D., S. M. Thomas, F. Niemitz, B. Reineking, and C. Beierkuhnlein. 2011b. Projection of climatic suitability for *Aedes albopictus* Skuse (Culicidae) in Europe under climate change conditions. Global and Planetary Change 78:54–64.

Fitzpatrick, M. C., J. F. Weltzin, N. J. Sanders, and R. R. Dunn. 2007. The biogeography of prediction error: Why does the introduced range of the fire ant overpredict its native range? Global Ecology and Biogeography 16:24–33.

Formenty, P., M. O. Muntasir, I. Damon, V. Chowdhary, M. L. Opoka, C. Monimart, E. M. Mutasim, J. C. Manuguerra, W. B. Davidson, K. L. Karem, J. Cabeza, S. Wang, M. R. Malik, T. Durand, A. Khalid, T. Rioton, A. Kuong-Ruay,

A. A. Babiker, M. E. Karsani, and M. S. Abdalla. 2010. Human monkeypox outbreak caused by novel virus belonging to Congo Basin clade, Sudan, 2005. Emerging Infectious Diseases 16:1539–1545.

Gage, K. L., and M. Y. Kosoy. 2005. The natural history of plague: Perspectives from more than a century of research. Annual Review of Entomology 50:505–528.

Gardner, L. M., D. Fajardo, S. T. Waller, O. Wang, and S. Sarkar. 2012. A predictive spatial model to quantify the risk of air-travel-associated dengue importation into the United States and Europe. Journal of Tropical Medicine 2012:103679.

Gilbert, M., X. Xiao, D. U. Pfeiffer, M. Epprecht, S. Boles, C. Czarnecki, P. Chaitaweesub, W. Kalpravidh, P. Q. Minh, M. J. Otte, V. Martin, and J. Slingenbergh. 2008. Mapping H5N1 highly pathogenic avian influenza risk in Southeast Asia. Proceedings of the National Academy of Sciences USA 105:4769–4774.

Giles, J., A. T. Peterson, and A. Almeida. 2010. Ecology and geography of plague transmission areas in northeastern Brazil. PLoS Neglected Tropical Diseases 5:e925.

Girard, J. M., D. M. Wagner, A. J. Vogler, C. Keys, C. J. Allender, L. C. Drickamer, and P. Keim. 2004. Differential plague-transmission dynamics determine *Yersinia pestis* population genetic structure on local, regional, and global scales. Proceedings of the National Academy of Sciences USA 101:8408–8413.

Glass, G. E., J. E. Cheek, J. A. Patz, T. M. Shields, T. J. Doyle, D. A. Thoroughman, D. K. Hunt, R. E. Enscore, K. L. Gage, C. Irland, C. J. Peters, and R. Bryan. 2000. Using remotely sensed data to identify areas at risk for hantavirus pulmonary syndrome. Emerging Infectious Diseases 6:238–247.

Godman, F. D., and O. Salvin. 1879–1915. Biologia Centrali-Americana [published in 215 parts by various authors]. Taylor and Francis, London.

Godsoe, W. 2010. "I can't define the niche but I know it when I see it": A formal link between statistical theory and the ecological niche. Oikos 119:53–60.

González, C., E. Rebollar, S. Ibáñez-Bernal, I. Becker, E. Martínez-Meyer, A. T. Peterson, and V. Sánchez-Cordero. 2011. Current knowledge on *Leishmania* vectors in Mexico: How species' geographic distributions relate to transmission areas. American Journal of Tropical Medicine and Hygiene 85:839–846.

González, C., O. Wang, S. E. Strutz, C. González-Salazar, V. Sánchez-Cordero, and S. Sarkar. 2010. Climate change and risk of leishmaniasis in North America: Predictions from ecological niche models of vector and reservoir species. PLoS Neglected Tropical Diseases 4:e585.

Green, R. H. 1971. A multivariate statistical approach to the Hutchinsonian niche: Bivalve molluscs of central Canada. Ecology 52:544–556.

Grinnell, J. 1914. Barriers to distribution as regards birds and mammals. American Naturalist 48:248–254.

Grinnell, J. 1917. Field tests of theories concerning distributional control. American Naturalist 51:115–128.

Grinnell, J. 1924. Geography and evolution. Ecology 5:225–229.

Grolla, A., A. Lucht, D. Dick, J. E. Strong, and H. Feldmann. 2005. Laboratory diagnosis of Ebola and Marburg hemorrhagic fever. Bulletin de la Société de Pathologie Exotique 98:205–209.

Grubesic, T. H., and T. C. Matisziw. 2006. On the use of ZIP codes and ZIP code tabulation areas (ZCTAs) for the spatial analysis of epidemiological data. International Journal of Health Geographics 5:58.

Guan, Y., B. J. Zheng, Y. Q. He, X. L. Liu, Z. X. Zhuang, and C. L. Cheung. 2003. Isolation and characterization of viruses related to the SARS coronavirus from animals in southern China. Science 302:276–278.

Guisan, A., J. Elith, C. H. Graham, S. J. Phillips, A. T. Peterson, and N. E. Zimmermann. 2007a. What matters for predicting spatial distributions of trees: Techniques, data, or species' characteristics? Ecological Monographs 77:615–630.

Guisan, A., C. H. Graham, J. Elith, F. Huettman, and NCEAS Species Distribution Modelling Group. 2007b. Sensitivity of predictive species distribution models to change in grain size. Diversity and Distributions 13:332–340.

Guralnick, R. P., H. Constable, J. Wieczorek, C. Moritz, and A. T. Peterson. 2009. Sharing: Lessons from natural history's success story. Nature 462:34.

Guralnick, R. P., and A. Hill. 2009. Biodiversity informatics: Automated approaches for documenting global biodiversity patterns and processes. Bioinformatics 25:421–428.

Guralnick, R. P., J. Wieczorek, R. Beaman, and R. J. Hijmans. 2006. BioGeomancer: Automated georeferencing to map the world's biodiversity data. PLoS Biology 4:e381.

Hales, S., N. de Wet, J. Maindonald, and A. Woodward. 2002. Potential effect of population and climate changes on global distribution of dengue fever: An empirical model. Lancet 360:830–834.

Hall, E. R. 1981. The Mammals of North America, 2nd ed. John Wiley and Sons, New York.

Hall-Baker, P. A., S. L. Groseclose, R. A. Jajosky, D. A. Adams, P. Sharp, W. J. Anderson, J. P. Abellera, A. E. Aranas, M. Mayes, M. S. Wodajo, D. H. Onweh, M. Park, and J. Ward. 2011. Summary of notifiable diseases—United States, 2009. Morbidity and Mortality Weekly Report 53:1–100.

Hamer, G. L., U. D. Kitron, J. D. Brawn, S. R. Loss, M. O. Ruiz, T. L. Goldberg, and E. D. Walker. 2008. *Culex pipiens* (Diptera: Culicidae): A bridge vector of West Nile virus to humans. Journal of Medical Entomology 45:125–128.

Harrigan, R. J., H. A. Thomassen, W. Buermann, R. F. Cummings, M. E. Kahn, and T. B. Smith. 2010. Economic conditions predict prevalence of West Nile virus. PLoS ONE 5:e15437.

Hay, S. I., K. E. Battle, D. M. Pigott, D. L. Smith, C. L. Moyes, S. Bhatt, J. S. Brownstein, N. Collier, M. F. Myers, D. B. George, and P. W. Gething. 2013. Global mapping of infectious disease. Philosophical Transactions of the Royal Society B 368:20120250.

Hay, S. I., J. Cox, D. J. Rogers, S. E. Randolph, D. I. Stern, G. D. Shanks, M. F. Myers, and R. W. Snow. 2002. Climate change and the resurgence of malaria in the East Africa highlands. Nature 415:905–909.

Heikkinen, R. K., M. Luoto, R. Virkkala, R. G. Pearson, and J.-H. Korber. 2007. Biotic interactions improve prediction of boreal bird distributions at macroscales. Global Ecology and Biogeography 16:754–763.

Hijmans, R., S. Cameron, J. Parra, P. Jones, and A. Jarvis. 2005. Very high resolution interpolated climate surfaces for global land areas. International Journal of Climatology 25:1965–1978.

Hirzel, A. H., J. Hausser, D. Chessel, and N. Perrin. 2002. Ecological-niche factor analysis: How to compute habitat-suitability maps without absence data? Ecology 83:2027–2036.

Ho, Y. C., and D. L. Pepyne. 2002. Simple explanation of the no-free-lunch theorem and its implications. Journal of Optimization Theory and Applications 115:549–570.

Hoen, A. G., G. Margos, S. J. Bent, M. A. Diuk-Wasser, A. Barbour, K. Kurtenbach, and D. Fish. 2009. Phylogeography of *Borrelia burgdorferi* in the eastern United States reflects multiple independent Lyme disease emergence events. Proceedings of the National Academy of Sciences USA 106:15013–15018.

Houck, M. A., H. Qin, and H. R. Roberts. 2001. Hantavirus transmission: Potential role of ectoparasites. Vector-Borne and Zoonotic Diseases 1:75–79.

Huete, A., K. Didan, T. Miura, E. P. Rodriguez, X. Gao, and L. G. Ferreira. 2002. Overview of the radiometric and biophysical performance of the MODIS vegetation indices. Remote Sensing of Environment 83:195–213.

Huey, R. B., and J. G. Kingsolver. 1993. Evolution of resistance to high temperature in ectotherms. American Naturalist 142:S21–S46.

Hutchinson, G. E. 1957. Concluding remarks. Cold Spring Harbor Symposia on Quantitative Biology 22:415–427.

Hutchinson, G. E. 1978. An Introduction to Population Ecology. Yale University Press, New Haven, CT.

IPCC [Intergovernmental Panel on Climate Change]. 2007. Climate Change 2007: The Physical Science Basis. Cambridge University Press, Cambridge.

ISRIC [International Soil Reference and Information Centre]. 2013. Soil property maps of Africa at 1 km. www.isric.org. ISRIC—World Soil Information, Wageningen, The Netherlands.

Iverson, L. R., A. Prasad, and M. W. Schwartz. 1999. Modeling potential future individual tree-species distributions in the eastern United States under a climate change scenario: A case study with *Pinus virginiana*. Ecological Modelling 115:77–93.

Jiménez-Valverde, A., N. Barve, A. Lira-Noriega, S. P. Maher, Y. Nakazawa, M. Papeş, J. Soberón, J. Sukumaran, and A. T. Peterson. 2010a. Dominant climate influences on North American bird distributions. Global Ecology and Biogeography 20:114–118.

Jiménez-Valverde, A., A. Lira-Noriega, A. T. Peterson, and J. Soberón. 2010b. Marshalling existing biodiversity data to evaluate biodiversity status and trends in planning exercises. Ecological Research 25:947–957.

Jiménez-Valverde, A., Y. Nakazawa, A. Lira-Noriega, and A. T. Peterson. 2009. Environmental correlation structure and ecological niche model projections. Biodiversity Informatics 6:28–35.

Johnson, P. T. J., and D. W. Thieltges. 2010. Diversity, decoys, and the dilution effect: How ecological communities affect disease risk. Journal of Experimental Biology 213:961–970.

Joly, D. O., M. D. Samuel, J. A. Langenberg, J. A. Blanchong, C. A. Batha, R. E. Rolley, D. P. Keane, and C. A. Ribic. 2006. Spatial epidemiology of chronic wasting disease in Wisconsin white-tailed deer. Journal of Wildlife Diseases 42:578–588.

Jones, K. E., N. G. Patel, M. A. Levy, A. Storeygard, D. Balk, J. L. Gittleman, and P. Daszak. 2008. Global trends in emerging infectious diseases. Nature 451: 990–993.

Jonsson, C. B., L. T. M. Figueiredo, and O. Vapalahti. 2010. A global perspective on hantavirus ecology, epidemiology, and disease. Clinical Microbiology Reviews 23:412–441.

Joyner, T. A., L. Lukhnova, Y. Pazilov, G. Temiralyeva, M. E. Hugh-Jones, A. Aikimbayev, and J. K. Blackburn. 2010. Modeling the potential distribution of *Bacillus anthracis* under multiple climate change scenarios for Kazakhstan. PLoS ONE 5:e9596.

Justice, C. O., E. Vermote, J. R. G. Townshend, R. Defries, D. P. Roy, D. K. Hall, V. V. Salomonson, J. L. Privette, G. Riggs, A. Strahler, W. Lucht, R. B. Myneni, Y. Knyazikhin, S. W. Running, R. R. Nemani, Z. Wan, A. R. Huete, W. van Leeuwen, R. E. Wolfe, L. Giglio, J.-P. Muller, P. Lewis, and M. J. Barnsley. 1998. The Moderate Resolution Imaging Spectroradiometer (MODIS): Land remote sensing for global change research. IEEE Transactions in Geosciences and Remote Sensing 36:1228–1249.

Kadmon, R., O. Farber, and A. Danin. 2004. Effect of roadside bias on the accuracy of predictive maps produced by bioclimatic models. Ecological Applications 14:401–413.

Kallio, E. R., J. Klingström, E. Gustafsson, T. Manni, A. Vaheri, H. Henttonen, O. Vapalahti, and Å. Lundkvist. 2006. Prolonged survival of Puumala hantavirus outside the host: Evidence for indirect transmission via the environment. Journal of General Virology 87:2127–2134.

Kearney, M. R., B. A. Wintle, and W. P. Porter. 2010. Correlative and mechanistic models of species distribution provide congruent forecasts under climate change. Conservation Letters 3:203–213.

Keeling, M. J., and P. Rohani. 2008. Modeling Infectious Diseases in Humans and Animals. Princeton University Press, Princeton, NJ.

Keesing, F., L. K. Belden, P. Daszak, A. Dobson, C. D. Harvell, R. D. Holt, P. Hudson, A. Jolles, K. E. Jones, C. E. Mitchell, S. S. Myers, T. Bogich, and R. S. Ostfeld. 2010. Impacts of biodiversity on the emergence and transmission of infectious diseases. Nature 468:647–652.

Keesing, F., R. D. Holt, and R. S. Ostfeld. 2006. Effects of species diversity on disease risk. Ecology Letters 9:485–498.

Kellermann, V., B. van Heerwaarden, C. M. Sgrò, and A. A. Hoffmann. 2009. Fundamental evolutionary limits in ecological traits drive *Drosophila* species distributions. Science 325:1244–1246.

Kilpatrick, A. M., A. A. Chmura, D. W. Gibbons, R. C. Fleischer, P. P. Marra, and P. Daszak. 2006. Predicting the global spread of H5N1 avian influenza. Proceedings of the National Academy of Sciences USA 103:19368–19373.

Kilpatrick, A. M., L. D. Kramer, S. R. Campbell, E. O. Alleyne, A. P. Dobson, and P. Daszak. 2005. West Nile virus risk assessment and the bridge vector paradigm. Emerging Infectious Diseases 11:425–429.

Koch, D. E., R. L. Mohler, and D. G. Goodin. 2007. Stratifying land use/land cover for spatial analysis of disease ecology and risk: An example using object-based classification techniques. Geospatial Health 2:15–28.

Koch, T. 2005. Cartographies of Disease: Maps, Mapping, and Medicine. ESRI Press, Redlands, CA.

Komar, N., R. S. Lanciotti, R. Bowen, S. Langevin, and M. Bunning. 2002. Detection of West Nile virus in oral and cloacal swabs collected from bird carcasses. Emerging Infectious Diseases 8:741–742.

Komar, N., S. Langevin, S. Hinten, N. Nemeth, E. Edwards, D. Hettler, B. Davis, R. Bowen, and M. Bunning. 2003. Experimental infection of North American birds with the New York 1999 strain of West Nile virus. Emerging Infectious Diseases 9:311–322.

Komar, O., M. B. Robbins, K. Klenk, B. J. Blitvich, N. L. Marlenee, K. L. Burkhalter, D. J. Gubler, G. Gonzálvez, C. J. Peña, A. T. Peterson, and N. Komar. 2003. West Nile virus transmission in resident birds, Dominican Republic. Emerging Infectious Diseases 9:1299–1302.

Kramer, L. D., L. M. Styer, and G. D. Ebel. 2008. A global perspective on the epidemiology of West Nile virus. Annual Review of Entomology 53:61–81.

Kulkarni, M. A., R. E. Desrochers, and J. T. Kerr. 2010. High resolution niche models of malaria vectors in northern Tanzania: A new capacity to predict malaria risk? PLoS ONE 5:e9396.

Kurane, I., and F. E. Ennis. 1992. Immunity and immunopathology in dengue virus infections. Seminars in Immunology 4:121–127.

Lacey, M. 2003. Nodding disease: Mystery of southern Sudan. Lancet Neurology 2:714.

Lai, P.-C., F.-M. So, and K.-W. Chan. 2009. Spatial Epidemiological Approaches in Disease Mapping and Analysis. CRC Press, Boca Raton, FL.

Lampe, K.-H., and D. Striebing. 2005. How to digitize large insect collections: Preliminary results of the DIG Project. Pages 385–393 in African Biodiversity: Molecules, Organisms, Ecosystems (B. A. Huber, B. J. Sinclair, and K.-H. Lampe, eds.). Springer, Dordrecht, The Netherlands.

Lanciotti, R. S., J. T. Roehrig, V. Deubel, J. Smith, M. Parker, K. Steele, B. Crise, K. E. Volpe, M. B. Crabtree, J. H. Scherret, R. A. Hall, J. S. MacKenzie, C. B. Cropp, B. Panigrahy, E. Ostlund, B. Schmitt, M. Malkinson, C. Banet, J. Weissman, N. Komar, H. M. Savage, W. Stone, T. McNamara, and D. J. Gubler. 1999. Origin of the West Nile virus responsible for an outbreak of encephalitis in the northeastern United States. Science 286:2333–2337.

LandScan. 2006. LandScan™ Global Population Database. www.ornl.gov/land scan/. Oak Ridge National Laboratory, Oak Ridge, TN.

Laporta, G. Z., D. G. Ramos, M. C. Ribeiro, and M. A. M. Sallum. 2011. Habitat suitability of Anopheles vector species and association with human malaria in the Atlantic Forest in south-eastern Brazil. Memórias do Instituto Oswaldo Cruz 106:239–245.

Lash, R. L., N. Brunsell, and A. T. Peterson. 2008. Spatiotemporal environmental triggers of Ebola and Marburg virus transmission. Geocarto International 23:451–466.

Lash R.R., D. S. Carroll, C. M. Hughes, Y. Nakazawa, K. Karem, I. K. Damon, and A. T. Peterson. 2012. Effects of georeferencing effort on mapping monkeypox case distributions and transmission risk. International Journal of Health Geographics 11:23.

Lawson, A. 2008. Bayesian Disease Mapping: Hierarchical Modeling in Spatial Epidemiology. Chapman and Hall, New York.

Lecompte, E., E. Fichet-Calvet, S. Daffis, K. Koulémou, O. Sylla, F. Kourouma, A. Doré, B. Soropogui, V. Aniskin, B. Allali, S. K. Kan, A. Lalis, L. Koivogui, S. Günther, C. Denys, and J. T. Meulen. 2006. *Mastomys natalensis* and Lassa fever, West Africa. Emerging Infectious Diseases 12:1971–1974.

Lee, C. E., J. L. Remfert, and G. W. Gelembiuk. 2003. Evolution of physiological tolerance and performance during freshwater invasions. Integrative and Comparative Biology 43:439–449.

Leroy, E. M., B. Kumulungui, X. Pourrut, P. Rouquet, A. Hassanin, P. Yaba, A. Delicat, J. T. Paweska, J. P. Gonzalez, and R. Swanepoel. 2005. Fruit bats as reservoirs of Ebola virus. Nature 438:575–576.

Levine, R. S., M. Q. Benedict, and A. T. Peterson. 2004a. Distribution of *Anopheles quadrimaculatus* Say *s.l.* and implications for its role in malaria transmission in the US. Journal of Medical Entomology 41:607–613.

Levine, R. S., A. T. Peterson, and M. Q. Benedict. 2004b. Geographic and ecologic distributions of the *Anopheles gambiae* complex predicted using a genetic algorithm. American Journal of Tropical Medicine and Hygiene 70:105–109.

Li, W., Z. Shi, M. Yu, W. Ren, C. Smith, J. H. Epstein, H. Wang, G. Crameri, Z. Hu, H. Zhang, J. Zhang, J. McEachern, H. Field, P. Daszak, B. T. Eaton, S. Zhang, and L.-F. Wang. 2005. Bats are natural reservoirs of SARS-like coronaviruses. Science 310:676–679.

Liu, C., P. Berry, T. Dawson, and R. Pearson. 2005. Selecting thresholds of occurrence in the prediction of species distributions. Ecography 28:385–393.

Lloyd-Smith, J. O., S. J. Schreiber, P. E. Kopp, and W. M. Getz. 2005. Superspreading and the effect of individual variation on disease emergence. Nature 438:355–359.

Lobo, J. M., A. Jiménez-Valverde, and R. Real. 2008. AUC: A misleading measure of the performance of predictive distribution models. Global Ecology and Biogeography 17:145–151.

Longstreth, J. D., and J. Wiseman. 1989. The potential impact of climate change on patterns of infectious disease in the United States. Appendix G: Health *in* The Potential Effects of Global Climate Change on the United States (J. B. Smith and D. A. Tirpak, eds.). U.S. Environmental Protection Agency, Washington, DC.

López-Cárdenas, J., F. E. González-Bravo, P. M. Salazar-Schettino, J. C. Gallaga-Solórzano, E. Ramírez-Barba, J. Martínez-Méndez, V. Sánchez-Cordero, A. T. Peterson, and J. M. Ramsey. 2005. Fine-scale predictions of distributions of Chagas disease vectors in the state of Guanajuato, Mexico. Journal of Medical Entomology 42:1068–1081.

MacArthur, R. H., and E. O. Wilson. 1967. The Theory of Island Biogeography. Princeton University Press, Princeton, NJ.

Machado-Machado, E. A. 2012. Empirical mapping of suitability to dengue fever in Mexico using species distribution modeling. Applied Geography 33:82–93.

Mackenzie, J. S., D. J. Gubler, and L. R. Petersen. 2004. Emerging flaviviruses: The spread and resurgence of Japanese encephalitis, West Nile, and dengue viruses. Nature Medicine 10:S98–S109.

Maguire, B. 1973. Niche response structure and the analytical potentials of its relationship to the habitat. American Naturalist 107:213–246.

Maher, S. P., C. Ellis, K. L. Gage, R. E. Enscore, and A. T. Peterson. 2010. Range-wide determinants of plague distribution in North America. American Journal of Tropical Medicine and Hygiene 83:736–742.

Marfin, A. A., and D. J. Gubler. 2001. West Nile encephalitis: An emerging disease in the United States. Clinical Infectious Diseases 33:1713–1719.

Marmion, M., M. Parviainen, M. Luoto, R. K. Heikkinen, and W. Thuiller. 2009. Evaluation of consensus methods in predictive species distribution modelling. Diversity and Distributions 15:59–69.

Martens, P., R. S. Kovats, S. Nijhof, P. de Vries, M. T. J. Livermore, D. J. Bradley, J. Cox, and A. J. McMichael. 1999. Climate change and future populations at risk of malaria. Global Environmental Change 9:S89–S107.

Martens, W. J. M., L. W. Niessen, and J. Rotmans. 1995. Potential impact of global climate change on malaria risk. Environmental Health Perspectives 103:458–464.

Martínez-Meyer, E. 2002. Evolutionary Trends in Ecological Niches of Species. PhD diss., University of Kansas.

Massafera, R., A. M. de Silva, A. P. de Carvalho, D. R. dos Santos, E. A. B. Galati, and U. Teodoro. 2005. Fauna de flebotomíneos do município de Bandeirantes, no estado do Paraná. Revista de Saúde Pública 39:571–577.

Masuoka, P., T. A. Klein, H. C. Kim, D. M. Claborn, N. Achee, R. Andre, J. Chamberlin, J. Small, A. Anyamba, D.-K. Lee, S. H. Yi, M. Sardelis, Y.-R. Ju, and J. Grieco. 2010. Modeling the distribution of *Culex tritaeniorhynchus* to predict Japanese encephalitis distribution in the Republic of Korea. Geospatial Health 5:45–57.

Maxcy, K. F. 1928. The distribution of endemic typhus (Brill's disease) in the United States. Public Health Reports 43:3084–3095.

Mayr, E. 1942. Systematics and the Origin of Species. Columbia University Press, New York.

McClish, D. 1989. Analyzing a portion of the ROC curve. Medical Decision Making 9:190–195.

McGarigal, K., S. A. Cushman, M. C. Neel, and E. Ene. 2002. FRAGSTATS: Spatial Pattern Analysis Program for Categorical Maps. www.umass.edu/landeco/research/fragstats/fragstats.html. University of Massachusetts, Amherst.

Milazzo, M. L., M. N. Cajimat, H. E. Romo, J. G. Estrada-Franco, L. I. Iñiguez-Dávalos, R. D. Bradley, and C. F. Fulhorst. 2012. Geographic distribution of hantaviruses associated with neotomine and sigmodontine rodents, Mexico. Emerging Infectious Diseases 18:571–576.

Mills, J. N., T. G. Ksiazek, C. J. Peters, and J. E. Childs. 1999. Long-term studies of hantavirus reservoir populations in the southwestern United States: A synthesis. Emerging Infectious Diseases 5:135–142.

Moffett, A., N. Shackelford, and S. Sarkar. 2007. Malaria in Africa: Vector species' niche models and relative risk maps. PLoS ONE 2:e824.

Monath, T. P. 2001. Yellow fever: An update. Lancet Infectious Diseases 1:11–20.

Monroe, M. C., S. P. Morzunov, A. M. Johnson, M. D. Bowen, H. Artsob, T. Yates, C. Peters, P. E. Rollin, T. G. Ksiazek, and S. T. Nichol. 1999. Genetic diversity and distribution of *Peromyscus*-borne hantaviruses in North America. Emerging Infectious Diseases 5:75–86.

Morales, M. A., M. Barrandeguy, C. Fabbri, J. B. Garcia, A. Vissani, K. Trono, G. Gutiérrez, S. Pigretti, H. Menchaca, N. Garrido, N. Taylor, F. Fernández, S. Levis, and D. Enría. 2006. West Nile virus isolation from equines in Argentina, 2006. Emerging Infectious Diseases 12:1559–1561.

Moyes, C. L., W. H. Temperley, A. J. Henry, C. R. Burgert, and S. I. Hay. 2013. Providing open access data online to advance malaria research and control. Malaria Journal 12:161.

Nakazawa, Y., R. Williams, A. T. Peterson, P. Mead, K. Kugeler, and J. Petersen. 2010. Ecological niche modeling of *Francisella tularensis* subspecies and clades in the United States. American Journal of Tropical Medicine and Hygiene 82:912–918.

Nakazawa, Y., R. Williams, A. T. Peterson, P. Mead, E. Staples, and K. L. Gage. 2007. Climate change effects on plague and tularemia in the United States. Vector-Borne and Zoonotic Diseases 7:529–540.

Navarro-Sigüenza, A. G., A. T. Peterson, and A. Gordillo-Martínez. 2003. Museums working together: The atlas of the birds of Mexico. Bulletin of the British Ornithologists' Club 123A:207–225.

Neerinckx, S. B., A. T. Peterson, H. Gulinck, J. Deckers, D. Kimaro, and H. Leirs. 2009a. Predicting potential risk areas of human plague for the Western Usambara Mountains, Lushoto District, Tanzania. American Journal of Tropical Medicine and Hygiene 82:492–500.

Neerinckx, S. B., A. T. Peterson, H. Gulinck, J. Deckers, and H. Leirs. 2009b. Geographic distribution and ecological niche of plague in Sub-Saharan Africa. International Journal of Health Geographics 7:54.

Negredo, A., G. Palacios, S. Vázquez-Morón, F. González, H. Dopazo, F. Molero, J. Juste, J. Quetglas, N. Savji, M. de la Cruz Martínez, J. E. Herrera, M. Pizarro, S. K. Hutchison, J. E. Echevarría, W. I. Lipkin, and A. Tenorio. 2011. Discovery of an Ebolavirus-like filovirus in Europe. PLoS Pathogens 7:e1002304.

Nelson, G. 1974. Historical biogeography: An alternative formalization. Systematic Zoology 23:555–558.

Nemirov, K., A. Vaheri, and A. Plyusnin. 2004. Hantaviruses: Co-evolution with natural hosts. Pages 201–228 *in* Recent Research Developments in Virology (S. G. Pandalai, ed.). Transworld Research Network, Trivandrum, India.

New, M., M. Hulme, and P. Jones. 1997. A 1961–1990 mean monthly climatology of global land areas. Climatic Research Unit, University of East Anglia, Norwich, UK.

Nix, H. A. 1986. A biogeographic analysis of Australian elapid snakes. Pages 4–15 *in* Atlas of Elapid Snakes of Australia (R. Longmore, ed.). Australian Government Publishing Service, Canberra.

NOAA [National Aeronautics and Space Administration]. 2012. VIIRS Nighttime Lights, 2012. National Geophysical Data Center, Boulder, CO.

Ogden, N. H., M. Bigras-Poulin, I. K. Barker, L. R. Lindsay, A. Maarouf, C. J. O'Callaghan, K. E. Smoyer-Tomic, D. Waltner-Toews, and D. Charron. 2005a. A dynamic population model to investigate effects of climate on geographic range and seasonality of the tick *Ixodes scapularis*. International Journal of Parasitology 35:375–389.

Ogden, N. H., A. Maarouf, I. K. Barker, M. Bigras-Poulin, L. R. Lindsay, M. G. Morshed, C. J. O'Callaghan, F. Ramay, D. Waltner-Toews, and D. F. Charron. 2005b. Climate change and the potential for range expansion of the Lyme disease vector *Ixodes scapularis* in Canada. International Journal for Parasitology 36: 63–70.

Ogden, N. H., L. St.-Onge, I. K. Barker, S. Brazeau, M. Bigras-Poulin, D. F. Charron, C. M. Francis, A. Heagy, L. R. Lindsay, A. Maarouf, P. Michel, F. Milord, C. J. O'Callaghan, L. Trudel, and R. A. Thompson. 2008. Risk maps for range expansion of the Lyme disease vector, *Ixodes scapularis*, in Canada now and with climate change. International Journal of Health Geographics 7:24.

Oliveira, C. C. G., H. G. Lacerda, D. R. M. Martins, J. D. A. Barbosa, G. R. Monteiro, J. W. Queiroz, J. M. A. Sousa, M. F. F. M. Ximenes, and S. M. B. Jeronimo. 2004. Changing epidemiology of American cutaneous leishmaniasis (ACL) in Brazil: A disease of the urban-rural interface. Acta Tropica 90:155–162.

Oliveira, D. M., K. R. Reinhold-Castro, M. V. Z. Bernal, C. M. de Oliveira Legriffon, M. V. C. Lonardoni, U. Teodoro, and T. G. V. Silveira. 2011. Natural infection of *Nyssomyia neivai* by *Leishmania* (*Viannia*) spp. in the state of Paraná, southern Brazil, detected by multiplex polymerase chain reaction. Vector-Borne and Zoonotic Diseases 11:137–143.

Openshaw, S. 1984. The Modifiable Areal Unit Problem. Geo Books, Norwich, UK.

Ortega-Huerta, M. A., and A. T. Peterson. 2008. Modeling ecological niches and predicting geographic distributions: A test of six presence-only methods. Revista Mexicana de la Biodiversidad 79:205–216.

Ostfeld, R. S., C. D. Canham, K. Oggenfuss, R. J. Winchcombe, and F. Keesing. 2006. Climate, deer, rodents, and acorns as determinants of variation in Lyme-disease risk. PLoS Biology 4:e145.

Ostfeld, R. S., and F. Keesing. 2000. Biodiversity and disease risk: The case of Lyme disease. Conservation Biology 14:722–728.

Owens, H. L., L. P. Campbell, L. Dornak, E. E. Saupe, N. Barve, J. Soberón, K. Ingenloff, A. Lira-Noriega, C. M. Hensz, C. E. Myers, and A. T. Peterson. 2013. Constraints on interpretation of ecological niche models by limited environmental ranges on calibration areas. Ecological Modelling 263:10–18.

Padula, P. J., A. Edelstein, S. D. L. Miguel, N. M. López, C. M. Rossi, and R. D. Rabinovich. 1998. Hantavirus pulmonary syndrome outbreak in Argentina: Molecular evidence for person-to-person transmission of Andes virus. Virology 241:323–330.

Parmenter, R. R., E. P. Yadev, C. A. Parmenter, P. Ettestad, and K. L. Gage. 1999. Incidence of plague associated with increased winter-spring precipitation in New Mexico, USA. American Journal of Tropical Medicine and Hygiene 61:814–821.

Pascual, M., J. A. Ahumada, L. F. Chaves, X. Rodó, and M. Bouma. 2006. Malaria resurgence in the East African highlands: Temperature trends revisited. Proceedings of the National Academy of Sciences USA 103:5829–5834.

Patz, J. A., W. J. M. Martens, D. A. Focks, and T. H. Jettend. 1998. Dengue fever epidemic potential as projected by general circulation models of global climate change. Environmental Health Perspectives 6:147–153.

Pearson, R. G., C. Raxworthy, M. Nakamura, and A. T. Peterson. 2007. Predicting species' distributions from small numbers of occurrence records: A test case using cryptic geckos in Madagascar. Journal of Biogeography 34:102–117.

Pearson, R. G., W. Thuiller, M. B. Araujo, E. Martínez-Meyer, L. Brotons, C. McClean, L. Miles, P. Segurado, T. P. Dawson, and D. C. Lees. 2006. Model-based uncertainty in species range prediction. Journal of Biogeography 33:1704–1711.

Petersen, L. R., and J. T. Roehrig. 2001. West Nile virus: A reemerging global pathogen. Emerging Infectious Diseases 7:1–10.

Peterson, A. T. 2003. Predicting the geography of species› invasions via ecological niche modeling. Quarterly Review of Biology 78:419–433.

Peterson, A. T. 2005. Predicting potential geographic distributions of invading species. Current Science 89:9.

Peterson, A. T. 2007a. Ecological niche modelling and understanding the geography of disease transmission. Veterinaria Italiana 43:393–400.

Peterson, A. T. 2007b. Why not WhyWhere: The need for more complex models of simpler environmental spaces. Ecological Modelling 203:527–530.

Peterson, A. T. 2008a. Biogeography of diseases: A framework for analysis. Naturwissenschaften 95:483–491.

Peterson, A. T. 2008b. Improving methods for reporting spatial epidemiologic data. Emerging Infectious Diseases 14:1335–1336.

Peterson, A. T. 2009. Shifting malaria transmission risk across Africa with warming climates. BMC Infectious Diseases 9:59.

Peterson, A. T. 2011. Ecological niche conservatism: A time-structured review of evidence. Journal of Biogeography 38:817–827.

Peterson, A. T. Forthcoming. Prototype system for tracking and forecasting highly pathogenic H5N1 avian influenza spread in North America. Studies in Avian Biology.

Peterson, A. T., M. J. Andersen, S. Bodbyl-Roels, P. Hosner, Á. Nyári, C. Oliveros, and M. Papeş. 2009. A prototype forecasting system for bird-borne disease spread in North America based on migratory bird movements. Epidemics 1:240–249.

Peterson, A. T., J. T. Bauer, and J. N. Mills. 2004a. Ecological and geographic distribution of filovirus disease. Emerging Infectious Diseases 10:40–47.

Peterson, A. T., B. W. Benz, and M. Papeş. 2007a. Highly pathogenic H5N1 avian influenza: Entry pathways into North America via bird migration. PLoS ONE 2:e261.

Peterson, A. T., D. Carroll, and J. N. Mills. 2004b. Potential mammalian filovirus reservoirs. Emerging Infectious Diseases 10:2073–2081.

Peterson, A. T., and M. T. Holder. 2012. Phylogenetic assessment of filoviruses: How many lineages of Marburg virus? Ecology and Evolution 2:1826–1833.

Peterson, A. T., S. Knapp, R. Guralnick, J. Soberón, and M. T. Holder. 2010. The big questions for biodiversity informatics. Systematics and Biodiversity 8:159–168.

Peterson, A. T., N. Komar, O. Komar, A. G. Navarro-Sigüenza, M. B. Robbins, and E. Martínez-Meyer. 2004c. West Nile virus in the New World: Potential impacts on bird species. Bird Conservation International 14:215–232.

Peterson, A. T., R. R. Lash, D. S. Carroll, and K. M. Johnson. 2006a. Geographic potential for outbreaks of Marburg hemorrhagic fever. American Journal of Tropical Medicine and Hygiene 75:9–15.

Peterson, A. T., C. Martínez-Campos, Y. Nakazawa, and E. Martínez-Meyer. 2005. Time-specific ecological niche modeling predicts spatial dynamics of vector insects and human dengue cases. Transactions of the Royal Society of Tropical Medicine and Hygiene 99:647–655.

Peterson, A. T., L. M. Moses, D. G. Bausch. 2014. Mapping transmission risk of Lassa fever in West Africa: The importance of quality control, sampling bias, and error weighting. PLoS ONE, forthcoming.

Peterson, A. T., and Y. Nakazawa. 2008. Environmental data sets matter in ecological niche modeling: An example with Solenopsis invicta and Solenopsis richteri. Global Ecology and Biogeography 17:135–144.

Peterson, A. T., and A. G. Navarro-Sigüenza. 2003. Computerizing bird collections and sharing collection data openly: Why bother? Bonner Zoologische Beiträge 51:205–212.

Peterson, A. T., and Á. Nyári. 2007. Ecological niche conservatism and Pleistocene refugia in the Thrush-like Mourner, Schiffornis sp., in the Neotropics. Evolution 62:173–183.

Peterson, A. T., M. A. Ortega-Huerta, J. Bartley, V. Sanchez-Cordero, J. Soberon, R. H. Buddemeier, and D. R. B. Stockwell. 2002a. Future projections for Mexican faunas under global climate change scenarios. Nature 416:626–629.

Peterson, A. T., M. Papeş, D. S. Carroll, H. Leirs, and K. M. Johnson. 2007b. Mammal taxa constituting potential coevolved reservoirs of filoviruses. Journal of Mammalogy 88:1544–1554.

Peterson, A. T., M. Papeş, and M. Eaton. 2007c. Transferability and model evaluation in ecological niche modeling: A comparison of GARP and Maxent. Ecography 30:550–560.

Peterson, A. T., M. Papeş, M. G. Reynolds, N. D. Perry, B. Hanson, R. L. Regnery, C. L. Hutson, I. K. Damon, and D. S. Carroll. 2006b. Invasive potential of Cricetomys rats in North America. Journal of Mammalogy 87:427–432.

Peterson, A. T., M. Papeş, and J. Soberón. 2008. Rethinking receiver operating characteristic analysis applications in ecological niche modelling. Ecological Modelling 213:63–72.

Peterson, A. T., R. S. Pereira, and V. L. Fonseca de Camargo-Neves. 2004d. Using epidemiological survey data to infer geographic distributions of Leishmania vector species. Revista da Sociedade Brasileira de Medicina Tropical 37:10–14.

Peterson, A. T., V. Sánchez-Cordero, C. B. Beard, and J. M. Ramsey. 2002b. Ecologic niche modeling and potential reservoirs for Chagas disease, Mexico. Emerging Infectious Diseases 8:662–667.

Peterson, A. T., V. Sánchez-Cordero, E. Martínez-Meyer, and A. G. Navarro-Sigüenza. 2006c. Tracking population extirpations via melding ecological niche modeling with land-cover information. Ecological Modelling 195:229–236.

Peterson, A. T., V. Sánchez-Cordero, J. Soberón, J. Bartley, R. H. Buddemeier, and A. G. Navarro-Sigüenza. 2001. Effects of global climate change on geographic distributions of Mexican Cracidae. Ecological Modelling 144:21–30.

Peterson, A. T., and J. J. Shaw. 2003. Lutzomyia vectors for cutaneous leishmaniasis in southern Brazil: Ecological niche models, predicted geographic distributions, and climate change effects. International Journal of Parasitology 33:919–931.

Peterson, A. T., J. Soberón, R. G. Pearson, R. P. Anderson, E. Martínez-Meyer, M. Nakamura, and M. B. Araújo. 2011. Ecological Niches and Geographic Distributions. Princeton University Press, Princeton, NJ.

Peterson, A. T., J. Soberón, and V. Sánchez-Cordero. 1999. Conservatism of ecological niches in evolutionary time. Science 285:1265–1267.

Peterson, A. T., D. A. Vieglais, and J. Andreasen. 2003. Migratory birds modeled as critical transport vectors for West Nile virus in North America. Vector-Borne and Zoonotic Diseases 3:39–50.

Peterson, A. T., and R. A. J. Williams. 2008. Risk mapping of highly pathogenic avian influenza distribution and spread. Ecology and Society 13:15, www.ecol ogyandsociety.org/vol13/iss2/art15/.

Petitpierre, B., C. Kueffer, O. Broennimann, C. Randin, C. Daehler, and A. Guisan. 2012. Climatic niche shifts are rare among terrestrial plant invaders. Science 335:1344–1348.

Pfeiffer, D., T. Robinson, M. Stevenson, K. Stevens, D. Rogers, and A. Clements. 2008. Spatial Analysis in Epidemiology. Oxford University Press, Oxford.

Phillips, S. J., R. P. Anderson, and R. E. Schapire. 2006. Maximum entropy modeling of species geographic distributions. Ecological Modelling 190:231–259.

Phillips, S. J., M. Dudik, J. Elith, C. Graham, A. Lehman, J. Leathwick, and S. Ferrier. 2009. Sample selection bias and presence-only distribution models: Implications for background and pseudo-absence data. Ecological Applications 19:181–197.

Pinzon, J. E., J. M. Wilson, C. J. Tucker, R. Arthur, P. B. Jahrling, and P. Formenty. 2004. Trigger events: Enviroclimatic coupling of Ebola hemorrhagic fever outbreaks. American Journal of Tropical Medicine and Hygiene 71:664–674.

Plowright, R. K., S. H. Sokolow, M. E. Gorman, P. Daszak, and J. E. Foley. 2008. Causal inference in disease ecology: Investigating ecological drivers of disease emergence. Frontiers in Ecology and the Environment 6:420–429.

Pulliam, H. R. 1988. Sources, sinks, and population regulation. American Naturalist 132:652–661.

Pulliam, H. R. 2000. On the relationship between niche and distribution. Ecology Letters 3:349–361.

Ramsey, J. M., A. Cruz-Celis, L. Salgado, L. Espinosa, R. Ordóñez, R. Lopez, and C. J. Schofield. 2003a. Efficacy of pyrethroid insecticides against domestic and peridomestic populations of Triatoma pallidipennis and Triatoma barberi (Reduviidae: Triatominae) vectors of Chagas' disease in Mexico. Journal of Medical Entomology 40:912–920.

Ramsey, J. M., R. Ordóñez, A. Tello-López, J. L. Pohls, V. Sánchez-Cordero, and A. T. Peterson. 2003b. Actualidades sobre la epidemiología de la enfermedad de Chagas en México. Pages 85–103 in Iniciativa para la Vigilancia y el Control de la Enfermedad de Chagas en la República Mexicana (J. M. Ramsey, A. Tello-López, and J. L. Pohls, eds.). Instituto Nacional de Salud Pública, Cuernavaca, Mexico.

Rappole, J., S. R. Derrickson, and Z. Hubálek. 2000. Migratory birds and spread of West Nile virus in the Western Hemisphere. Emerging Infectious Diseases 6:319–328.

Rappole, J., and Z. Hubálek. 2003. Migratory birds and West Nile virus. Journal of Applied Microbiology 94:47–58.

Renner, I. W., and D. I. Warton. 2013. Equivalence of MAXENT and Poisson point process models for species distribution modeling in ecology. Biometrics 69:274–281.

Riesenfeld, C. S., P. D. Schloss, and J. Handelsman. 2004. Metagenomics: Genomic analysis of microbial communities. Annual Review of Genetics 38:525–552.

Rios, N., and H. L. Bart. 2008. Community building and collaborative georeferencing using GEOLocate. Page 46 *in* The Proceedings of TDWG: Provisional Abstracts of the 2008 Annual Conference of the Taxonomic Databases Working Group (A. Weitzman and L. Belbin, eds.). Biodiversity Information Standards (TDWG) and Missouri Botanical Garden, Freemantle, Australia.

Robertson, M. P., N. Caithness, and M. H. Villet. 2001. A PCA-based modelling technique for predicting environmental suitability for organisms from presence records. Diversity and Distributions 7:15–27.

Rojas-Soto, O. R., O. Alcantara-Ayala, and A. G. Navarro. 2003. Regionalization of the avifauna of the Baja California peninsula, Mexico: A parsimony analysis of endemicity and distributional modelling approach. Journal of Biogeography 30:449–461.

Rose, H., and R. Wall. 2011. Modelling the impact of climate change on spatial patterns of disease risk: Sheep blowfly strike by *Lucilia sericata* in Great Britain. International Journal of Parasitology 41:739–746.

Royle, J. A., R. B. Chandler, C. Yackulic, and J. D. Nichols. 2012. Likelihood analysis of species occurrence probability from presence-only data for modelling species distributions. Methods in Ecology and Evolution 3:545–554.

Running, S. W., C. Justice, V. Salomonson, D. Hall, J. Barker, Y. Kaufman, A. Strahler, A. Huete, J. P. Muller, V. Vanderbilt, Z. M. Wan, P. Teillet, and D. Carneggie. 1994. Terrestrial remote sensing science and algorithms planned for EOS/MODIS. International Journal of Remote Sensing 15:3587–3620.

Ryckman, R. E. 1962. Biosystematics and hosts of the *Triatoma protracta* complex in North America (Hemiptera: Reduviidae) (Rodentia: Cricetidae). University of California Publications in Entomology 27:93–240.

Salkeld, D. J., and P. Stapp. 2006. Seroprevalence rates and transmission of plague (*Yersinia pestis*) in mammalian carnivores. Vector-Borne and Zoonotic Diseases 6:231–239.

Sánchez-Cordero, V., A. T. Peterson, E. Martínez-Meyer, and R. Flores. 2005. Distribución de roedores reservorios del virus causante del sindrome pulmonar por hantavirus y regiones de posible riesgo en México. Acta Zoológica Mexicana 21:79–91.

Sardelis, M. R., M. J. Turell, D. J. Dohm, and M. L. O'Guinn. 2011. Vector competence of selected North American *Culex* and *Coquillettidia* mosquitoes for West Nile virus. Emerging Infectious Diseases 7:1018–1022.

Sarkar, S., R. L. Pressey, D. P. Faith, C. R. Margules, T. Fuller, D. M. Stoms, A. Moffett, K. A. Wilson, K. J. Williams, P. H. Williams, and S. Andelman. 2006. Biodiversity conservation planning tools: Present status and challenges for the future. Annual Review of Environment and Resources 31:123–159.

Sastre, P., and J. M. Lobo. 2009. Taxonomist survey biases and the unveiling of biodiversity patterns. Biological Conservation 142:462–467.

Saupe, E., V. Barve, C. Myers, J. Soberón, N. Barve, C. Hensz, A. T. Peterson, H. L. Owens, and A. Lira-Noriega. 2012. Variation in niche and distribution model performance: The need for a priori assessment of key causal factors. Ecological Modelling 237–238:11–22.

Scharlemann, J. P. W., D. Benz, S. I. Hay, B. V. Purse, A. J. Tatem, G. R. W. Wint, and D. J. Rogers. 2008. Global data for ecology and epidemiology: A novel algorithm for temporal Fourier processing MODIS data. PLoS ONE 3:e1408.

Schmidt, K. A., and R. S. Ostfeld. 2001. Biodiversity and the dilution effect in disease ecology. Ecology 82:609–619.

Seaman, V. 1804. An inquiry into the cause of the prevalence of the yellow fever in New York. Medical Repository, 3rd ed., 1:303–323.

Si, Y., T. Wang, A. K. Skidmore, W. F. de Boer, L. Li, and H. H. T. Prins. 2010. Environmental factors influencing the spread of the highly pathogenic avian influenza H5N1 virus in wild birds in Europe. Ecology and Society 15:26.

Silva, A. M. de, N. J. de Camargo, D. R. dos Santos, R. Massafera, A. C. FerreiraI, C. Postai, E. C. Cristóvão, J. F. Konolsaisen, A. Bisetto Jr., R. Perinazo, U. Teodoro, and E. A. B. Galati. 2008. Diversidade, distribuição e abundância de flebotomíneos (Diptera: Psychodidae) no Paraná. Neotropical Entomology 37:209–225.

Simard, M., N. Pinto, J. B. Fisher, and A. Baccini. 2011. Mapping forest canopy height globally with space-borne LIDAR. Journal of Geophysical Research 116: G04021.

Siqueira, M. F. de, G. Durigan, P. de Marco Júnior, and A. T. Peterson. 2009. Something from nothing: Using landscape similarity and ecological niche modeling to find rare plant species. Journal for Nature Conservation 17:25–32.

Slater, H., and E. Michael. 2012. Predicting the current and future potential distributions of lymphatic filariasis in Africa using maximum entropy ecological niche modelling. PLoS ONE 7:e32202.

Smith, C. H., and G. Beccaloni. 2008. Natural Selection and Beyond: The Intellectual Legacy of Alfred Russel Wallace. Oxford University Press, Oxford.

Snow, J. 1855. On the Mode of Communication of Cholera, 2nd ed., much enlarged. John Churchill, London.

Soberón, J. 1999. Linking biodiversity information sources. Trends in Ecology and Evolution 14:291.

Soberón, J. 2007. Grinnellian and Eltonian niches and geographic distributions of species. Ecology Letters 10:1115–1123.

Soberón, J., and A. T. Peterson. 2005. Interpretation of models of fundamental ecological niches and species' distributional areas. Biodiversity Informatics 2:1–10.

Soberón, J., and A. T. Peterson. 2009. Monitoring biodiversity loss with primary species-occurrence data: Toward national-level indicators for the 2010 Target of the Convention on Biological Diversity. Ambio 38:29–34.

Soberón, J., and A. T. Peterson. 2011. Ecological niche shifts and environmental space anisotropy: A cautionary note. Revista Mexicana de Biodiversidad 82:1348–1353.

Squires, R. B., J. Noronha, V. Hunt, A. García-Sastre, C. Macken, N. Baumgarth, D. Suarez, B. E. Pickett, Y. Zhang, C. N. Larsen, A. Ramsey, L. Zhou, S. Zaremba, S. Kumar, J. Deitrich, E. Klem, and R. H. Scheuermann. 2012. Influenza Research Database: An integrated bioinformatics resource for influenza research and surveillance. Influenza and Other Respiratory Viruses 6:404–416.

Stein, B. R., and J. Wieczorek. 2004. Mammals of the world: MaNIS as an example of data integration in a distributed network environment. Biodiversity Informatics 1:14–22.

Stevens, K. B., and D. U. Pfeiffer. 2011. Spatial modelling of disease using data- and knowledge-driven approaches. Spatial and Spatio-Temporal Epidemiology 2:125–133.

Stockwell, D. R. B., J. H. Beach, A. Stewart, G. Vorontsov, D. Vieglais, and R. S. Pereira. 2006. The use of the GARP genetic algorithm and Internet grid computing in the Lifemapper world atlas of species biodiversity. Ecological Modelling 195:139–145.

Stockwell, D. R. B., and D. P. Peters. 1999. The GARP modelling system: Problems and solutions to automated spatial prediction. International Journal of Geographical Information Science 13:143–158.

Storfer, A., M. A. Murphy, J. S. Evans, C. S. Goldberg, S. Robinson, S. F. Spear, R. Dezzani, E. Delmelle, L. Vierling, and L. P. Waits. 2006. Putting the "landscape" in landscape genetics. Heredity 98:128–142.

Sucaet, Y., J. Van Hemert, B. Tucker, and L. Bartholomay. 2008. A web-based relational database for monitoring and analyzing mosquito population dynamics. Journal of Medical Entomology 45:775–784.

Suzán, G., G. Ceballos, J. Mills, T. G. Ksiazek, and T. Yates. 2001. Serologic evidence of hantavirus infection in sigmodontine rodents in Mexico. Journal of Wildlife Diseases 37:391–393.

Swanepoel, R., S. B. Smit, P. E. Rollin, P. Formenty, P. A. Leman, A. Kemp, F. J. Burt, A. A. Grobbelaar, J. Croft, D. G. Bausch, H. Zeller, H. Leirs, L. Braack, M. L. Libande, S. Zaki, S. T. Nichol, T. G. Ksiazek, and J. T. Paweska. 2007. Studies of reservoir hosts for Marburg virus. Emerging Infectious Diseases 13:1847–1851.

Sweeney, A. W., N. W. Beebe, R. D. Cooper, J. T. Bauer, and A. T. Peterson. 2006. Environmental factors associated with the distribution and range limits of the malaria vector *Anopheles farauti sensu stricto* in Australia. Journal of Medical Entomology 43:1068–1075.

Tabachnick, W. J. 1991. Evolutionary genetics and arthropod-borne disease: The yellow fever mosquito. American Entomologist 37:14–26.

Talani, P., J. Konongo, A. Gromyko, J. Nanga-Maniane, F. Yala, and D. Bodzongo. 1999. Prévalence des anticorps anti-fièvres hémorragiques d'origine virale dans la région du Pool (Congo-Brazzaville). Médecine d'Afrique Noire 46:424–427.

Taniguchi, S., S. Watanabe, J. S. Masangkay, T. Omatsu, T. Ikegami, P. Alviola, N. Ueda, K. Iha, H. Fujii, Y. Ishii, T. Mizutani, S. Fukushi, M. Saijo, I. Kurane, S. Kyuwa, H. Akashi, Y. Yoshikawa, and S. Morikawa. 2011. Reston Ebolavirus antibodies in bats, the Philippines. Emerging Infectious Diseases 17:1559–1560.

Tanser, F. C., B. Sharp, and D. le Sueur. 2003. Potential effect of climate change on malaria transmission in Africa. Lancet 362:1792–1798.

Taylor, R. M., T. H. Work, H. S. Hurlbut, and F. Rizk. 1956. A study of the ecology of West Nile virus in Egypt. American Journal of Tropical Medicine and Hygiene 5:579–620.

TDWG [Taxonomic Databases Working Group]. 2007a. Access to Biological Collections Data—ABCD. www.tdwg.org/activities/abcd/. Biodiversity Information Standards—TDWG.

TDWG. 2007b. DarwinCore Group—DwC. www.tdwg.org/activities/darwincore/. Biodiversity Information Standards—TDWG.

Tesh, R. B., A. P. A. Travassos da Rosa, H. Guzman, T. P. Araújo, and S.-Y. Xiao. 2002. Immunization with heterologous flaviviruses protective against fatal West Nile encephalitis. Emerging Infectious Diseases 8:245–251.

Thomson, M. C., D. A. Elnaiem, R. W. Ashford, and S. J. Connor. 1999. Towards a kala-azar risk map for Sudan: Mapping the potential distribution of *Phlebotomus orientalis* using digital data of environmental variables. Tropical Medicine and International Health 4:105–113.

Thuiller, W., M. B. Araújo, and S. Lavorel. 2004. Do we need land-cover data to model species distributions in Europe? Journal of Biogeography 31:353–361.

Towner, J. S., B. R. Amman, T. K. Sealy, S. A. R. Carroll, J. A. Comer, A. Kemp, R. Swanepoel, C. D. Paddock, S. Balinandi, M. L. Khristova, P. B. H. Formenty, C. G. Albarino, D. M. Miller, Z. D. Reed, J. T. Kayiwa, J. N. Mills, D. L. Cannon, P. W. Greer, E. Byaruhanga, E. C. Farnon, P. Atimnedi, S. Okware, E. Katongole-Mbidde, R. Downing, J. W. Tappero, S. R. Zaki, T. G. Ksiazek, S. T. Nichol, and P. E. Rollin. 2009. Isolation of genetically diverse Marburg viruses from Egyptian fruit bats. PLoS Pathogens 5:e1000536.

Towner, J. S., X. Pourrut, C. G. Albariño, C. N. Nkogue, B. H. Bird, G. Grard, T. G. Ksiazek, J.-P. Gonzalez, S. T. Nichol, and E. M. Leroy. 2007. Marburg virus infection detected in a common African bat. PLoS ONE 8:e764.

Turell, M. J., D. J. Dohm, M. R. Sardelis, M. L. O'Guinn, T. G. Andreadis, and J. A. Blow. 2005. An update on the potential of North American mosquitoes (Diptera: Culicidae) to transmit West Nile virus. Journal of Medical Entomology 42:57–62.

van de Groen, G., K. M. Johnson, F. A. Webb, H. Wulff, and J. Lange. 1978. Results of Ebola antibody surveys in various population groups. Pages 203–205 *in* Ebola Virus Haemorrhagic Fever (S. R. Pattyn, ed.). Elsevier / North Holland Biomedical Press, Amsterdam.

Vitek, C. R., R. F. Breiman, T. G. Ksiazek, P. E. Rollin, J. C. McLaughlin, E. T. Umland, K. B. Nolte, A. Loera, C. M. Sewell, and C. J. Peters. 1996. Evidence against person-to-person transmission of hantavirus to health care workers. Clinical Infectious Diseases 22:824–826.

Wallace, A. R. 1860. On the zoological geography of the Malay Archipelago. Proceedings of the Linnean Society of London 4:172–184.

Waller, L. A., B. Goodwin, M. Wilson, R. Ostfeld, S. Marshall, and E. Hayes. 2007. Spatio-temporal patterns in county-level incidence and reporting of Lyme disease in the northeastern United States, 1990–2000. Environmental and Ecological Statistics 14:83–100.

Waller, L. A., and C. A. Gotway. 2004. Applied Spatial Analysis of Public Health Data. John Wiley and Sons, Hoboken, NJ.

Ward, G., T. Hastie, S. Barry, J. Elith, and J. R. Leathwick. 2009. Presence-only data and the EM algorithm. Biometrics 65:554–563.

Warren, D. L., R. E. Glor, and M. Turelli. 2008. Environmental niche equivalency versus conservatism: Quantitative approaches to niche evolution. Evolution 62:2868–2883.

Warren, D. L., and S. N. Seifert. 2011. Ecological niche modeling in Maxent: The importance of model complexity and the performance of model selection criteria. Ecological Applications 21:335–342.

Webb, C. T., C. P. Brooks, K. L. Gage, and M. F. Antolin. 2006. Classic flea-borne transmission does not drive plague epizootics in prairie dogs. Proceedings of the National Academy of Sciences USA 103:6236–6241.

Weinstein, R. A., C. B. Bridges, M. J. Kuehnert, and C. B. Hall. 2003. Transmission of influenza: Implications for control in health care settings. Clinical Infectious Diseases 37:1094–1101.

Wells, R. M., S. S. Estani, Z. E. Yadon, D. Enria, P. J. Padula, N. Pini, J. N. Mills, C. J. Peters, and E. L. Segura. 1997a. An unusual hantavirus outbreak in southern Argentina: Person-to-person transmission? Emerging Infectious Diseases 3:171–174.

Wells, R. M., J. Young, R. J. Williams, L. R. Armstrong, K. Busico, A. S. Khan, T. G. Ksiazek, P. E. Rollin, S. R. Zaki, S. T. Nichol, and C. J. Peters. 1997b. Hantavirus transmission in the United States. Emerging Infectious Diseases 3:361–365.

Wieczorek, J. 2001. MaNIS Georeferencing Calculator, Vol. 2011. Museum of Vertebrate Zoology, University of California, Berkeley.

Wieczorek, J., Q. Guo, and R. Hijmans. 2004. The point-radius method for georeferencing locality descriptions and calculating associated uncertainty. International Journal of Geographical Information Science 18:745–767.

Wiley, E. O. 1981. The Theory and Practice of Phylogenetic Systematics, 1st ed. John Wiley and Sons, New York.

Williams, P., D. Gibbons, C. R. Margules, A. Rebelo, C. Humphries, and R. Pressey. 1996. A comparison of richness hotspots, rarity hotspots, and complementary areas for conserving diversity of British birds. Conservation Biology 10:155–174.

Williams, R. A. J., F. O. Fasina, and A. T. Peterson. 2008. Predictable ecology and geography of avian influenza (H5N1) transmission in Nigeria and West Africa. Transactions of the Royal Society of Tropical Medicine and Hygiene 102:471–479.

Williams, R. A. J., and A. T. Peterson. 2009. Ecology and geography of avian influenza (HPAI H5N1) transmission in the Middle East and northeastern Africa. International Journal of Health Geographics 8:47.

Williams, R. A. J., X. Xiao, and A. T. Peterson. 2011. Continent-wide association of H5N1 outbreaks in wild and domestic birds in Europe. Geospatial Health 5:247–253.

Williamson, H. R., M. E. Benbow, L. P. Campbell, C. R. Johnson, G. Sopoh, Y. Barogui, R. W. Merritt, and P. L. C. Small. 2012. Detection of *Mycobacterium ulcerans* in the environment predicts prevalence of Buruli ulcer in Benin. PLoS Neglected Tropical Diseases 6:e1506.

Winkler, A. S., K. Friedrich, R. König, M. Meindl, R. Helbok, I. Unterberger, T. Gotwald, J. Dharsee, S. Velicheti, A. Kidunda, L. Jilek-Aall, W. Matuja, and E. Schmutzhard. 2008. The head nodding syndrome: Clinical classification and possible causes. Epilepsia 49:2008–2015.

Wonham, M. J., M. A. Lewis, J. Rencławowicz, and P. van den Driessche. 2006. Transmission assumptions generate conflicting predictions in host-vector disease models: A case study in West Nile virus. Ecology Letters 9:706–725.

Work, T. H., H. S. Hurlbut, and R. M. Taylor. 1955. Indigenous wild birds of the Nile delta as potential West Nile virus circulating reservoirs. American Journal of Tropical Medicine and Hygiene 4:872–888.

Xavier, S. C. das C., A. L. R. Roque, V. dos S. Lima, K. J. L. Monteiro, J. C. R. Otaviano, L. F. C. F. da Silva, and A. M. Jansen. 2012. Lower richness of small wild mammal species and Chagas disease risk. PLoS Neglected Tropical Diseases 6:e1647.

Yanagihara, R. 1990. Hantavirus infection in the United States: Epizootiology and epidemiology. Review of Infectious Diseases 12:449–457.

Yee, K. S., T. E. Carpenter, and C. J. Cardona. 2009. Epidemiology of H5N1 avian influenza. Comparative Immunology, Microbiology and Infectious Diseases 32:325–340.

Yesson, C., P. W. Brewer, T. Sutton, N. Caithness, J. S. Pahwa, M. Burgess, W. A. Gray, R. J. White, A. C. Jones, F. A. Bisby, and A. Culham. 2007. How global is the Global Biodiversity Information Facility? PLoS ONE 2:e1124.

Zeller, H. G., and I. Schuffenecker. 2004. West Nile virus: An overview of its spread in Europe and the Mediterranean basin in contrast to its spread in the Americas. European Journal of Clinical Microbiology and Infectious Diseases 23:147–156.

Zendejas-Martínez, H., A. T. Peterson, and F. Milián-Suazo. 2008. Coarse-scale spatial and ecological analysis of tuberculosis in cattle: An investigation in Jalisco, Mexico. Geospatial Health 3:29–38.

Zurell, D., J. Elith, and B. Schröder. 2012. Predicting to new environments: Tools for visualizing model behaviour and impacts on mapped distributions. Diversity and Distributions 18:628–634.

Index

absence, 33–35, 134–36
accessible area, 11, 16, 81–83, 92–93, 105–6, 126–27
algorithms, 116–17
area under the curve, 131–32, 136–38
autocorrelation, 97–98, 133

BAM, 8, 11–13, 13–14, 33–35, 104–9
bias, 79–80, 81
biotic interactions, 11, 12, 14–15, 143
black-box approaches, 60–63, 142, 154–55, 159–60, 183

calibration, 113–23
climate change, 171–79
climate data, 100
compatibility, 92–93, 99
complexity of models, 119
components of transmission systems, 53–58, 63–67, 142, 155–56, 161, 183
confusion matrix, 134–36

data cleaning, 90–92
data quality, 90–92
data splitting, 117–18, 132–33
data subsetting, 117–18, 132–33
dimensionality, 98–99, 119
disease transmission risk, 3, 48–49, 124, 140–49, 153–57
dispersal, 127–28, 167–69

Eltonian Noise Hypothesis, 47–48
environmental data, 94–103
error, 75–83, 90–92, 118–19, 148
evaluation, 130–39

existing fundamental ecological niche, 12, 122
exposure, 87
extrapolation, 121–22, 127

fundamental ecological niche, 9–10, 12–13

georeferencing, 28–29, 30–33, 86–88, 89–90
Grinnell, Joseph, 2, 7–8

human-related data, 102–3, 143–48, 156
Hutchinson, G. Evelyn, 8, 11

interpolation, 38–39, 68–72, 95–96, 158–63

kappa statistic, 41–42, 135

M, 11, 16, 81–83, 92–93, 105–6, 126–27
model calibration, 113–23
model complexity, 119
model evaluation, 130–39
model performance, 138
model thresholding, 125–26, 136–38
model transfer, 71, 108–9, 121–22, 127
model validation, 130–39
Modifiable Areal Unit Problem, 99–100
movement, 127–28, 167–69

niche, 7–10
niche conservatism, 58–60
no-free-lunch theorem, 119–21
no-silver-bullet hypothesis, 119–21

occurrence data, 71, 84–93
overfitting, 98–99, 119